FIRST BOOK

OF

ZOÖLOGY.

BY

EDWARD S. MORSE, Ph. D.,

LATE PROFESSOR OF COMPARATIVE ANATOMY AND ZOÖLOGY IN BOWDOIN COLLEGE.

"As for your pretty little seed-cups, or vases, they are a sweet confirmation of the pleasure Nature seems to take in superadding an elegance of form to most of her works, wherever you find them. How poor and bungling are all the imitations of art! When I have the pleasure of seeing you next we will sit down—nay, kneel down if you will—and admire these things." —[Hogarth *in a Letter to* Ellis.

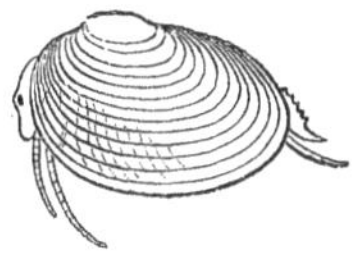

PREFACE.

THE "First Book of Zoölogy" is expressly prepared for the use of pupils who wish to gain a general knowledge of the structure, habits, modes of growth, and other leading features concerning the common animals of the country.

Particular attention has been given to the lower animals, as these are more often neglected in text-books. Directions for collecting, the preparation of specimens for the cabinet, and the haunts of the animals to be studied are given, and the pupil is expected to study, with the book in one hand, and the specimens in the other. The figures illustrating this work, with a few exceptions, have been drawn from Nature by the author, and have been prepared with especial reference to their being copied by the pupil. Tc facilitate this the figures are made in outline, with the shaded side of the figure indicated by darker lines.

The necessity of the pupils copying (however poorly) the figures, either upon the slate, or upon paper, cannot be too strongly urged.

From his own experience, the author has learned that a specimen or figure may oftentimes be carefully studied, and yet only an imperfect idea be formed of it; but, when it had been once copied, the new points gained repaid all the trouble spent in the task.

It makes but little difference whether the pupil is proficient in drawing or not; it should be strenuously insisted upon by the teacher that the pupils copy, as far as possible, the figures contained in each lesson.

To collect in the field, to make a cabinet, and then to examine and study the specimens collected, are the three stages that naturalists, with few exceptions, have passed through in their boyhood.

If one recalls the way in which boys first manifest their taste for such studies, he will remember that first a few examples were brought together; a collection was made. It may have been birds'-eggs, insects, or shells; then little boxes, a case of drawers, or shelves, were secured to hold their treasures. In thus collecting and arranging and rearranging the cabinet, the eye becomes familiar with the outline and general character of the objects, and in this way the mind is finally prepared to comprehend the relations existing between animals, and to appreciate the leading points upon which classification is founded.

Agassiz invariably placed before his students a single specimen, or a box full of specimens, and told them to look and see what they could find out.

It has seemed, therefore, that the way to commence the study of zoölogy is to follow the course one naturally pursues when he is led to the study by predisposition. Nor is it essential, at the outset, to present the entire range of the animal kingdom. Teach the characters of one or two great divisions first, and then the pupil is better prepared to grasp in turn the other divisions. The persistent attempt, in all text-books of this kind, to give some attention to every large group in the animal kingdom, has often resulted in wearying and confusing the minds of those who take up the study for the first time.

A very serious difficulty is encountered in those books which give a more or less complete view of systematic zoölogy for beginners. In some, the authors commence with the lowest forms, and end with the highest. In others, the highest animals are dealt with first, and the lessons end with the lowest. The first mistake made is the attempt to teach systematic zoölogy, where the pupil is quite ignorant of the material to be classified; and proper familiarity with the objects, the author contends, can only be acquired by collecting the specimens and forming a little cabinet of them.

The difficulty, however, arises in commencing the lessons with either the lowest or the highest animals. If the author commences with the lowest animals, he deals at the outset with creatures which the pupil in certain cases can never see, as many of the animals to be considered are microscopic, and most of them of such a nature that their soft parts cannot be preserved. On the other hand, if the author commence with the vertebrates, he presents, point-blank, some of the forms of structure most difficult to understand.

The main thing at the outset is to teach the pupil how to collect the objects for study; this leads him to observe them in Nature, and here the best part of the lesson is learned: methods of protection for the young, curious habits, modes of fabricating nests, and many little features are here observed, which can never be studied from an ordinary collection. Hence, collecting in the field is of paramount importance. Next, the forming of a little collection at home prompts the pupil to seek out certain resemblances among his objects, in order to bring those of a kind together. In this way he is prepared to understand and appreciate methods of classification. Finally, having grasped the leading features of a few groups, he is enabled to comprehend the character of cognate groups with less difficulty. Thus, an inland student, having got the typical idea of an insect from the study of a common grasshopper, for example, is much better prepared to understand the general structure of the Crustacea, though he may never have seen the few forms peculiar to fresh water. In the same way after having studied the common earthworm, he can form a better idea of the complicated structure of many marine worms, though these he may never see. After long deliberation, and some hesitancy, the author is forced to depart from common usage, and present, in this first book, only a few of the leading groups in the animal kingdom.

From the abundance of material, and the comparative ease with which the specimens may be preserved for cabinet use, shells and insects have always formed the favorite collections of children. They are the most common objects in nearly all collections, and it has seemed to the author that here the pupil ought to commence his studies.

Having learned to collect and prepare specimens for the cabinet, and to observe the relations and differences existing among them, the pupil is then prepared to go on to forms less familiar, or to study in detail the material already gone over.

Great pains have been taken to present, in every case, drawings made from the animal, expressly for this book. They are all American, and, with few exceptions, are entirely new. It is believed that teachers will appreciate the absence of those hackneyed illustrations which have too long done service in text-books on the subject.

To those especially interested in the study, many figures are given which have never before been published, even in scientific works, as, for example, the rare *Lymnæa megasoma*, *Lymnæa ampla*, *Ptyelus lineatus*, and many others.

I desire here to express my thanks to Dr. A. S. Packard, for looking over the pages relating to insects, and to Dr. H. Hagen, Mr. Samuel H. Scudder, Prof. H. H. Straight, Prof. Theodore Gill, Prof. A. J. Cook, and Miss Maggie W. Brooks, for important specimens for illustration; and to the firm of Russell & Richardson, Boston, who, with the interest of personal friends, have attended to the proper engraving of my drawings. I am also deeply indebted to Prof. E. L. Youmans, Mr. John M. Gould, and Mr. Henry W. Swasey, for valuable suggestions and advice; and, finally, I have to express my gratitude to the publishers of this book, who have, with unbounded liberality, left the entire matter of illustration in my hands.

E. S. M.

SALEM, MASS., *March* 12, 1875.

CONTENTS.

CHAPTER VI.

CHAPTER VII.

CHAPTER VIII.

CHAPTER IX.

CHAPTER X.

CHAPTER XI.

CHAPTER XII.

CHAPTER XIII.

CHAPTER XIV.

CHAPTER XV.

CHAPTER XVI.

CHAPTER XVII.

CHAPTER XVIII.

CHAPTER XIX.

FIRST BOOK OF ZOÖLOGY.

CHAPTER I.

FRESH-WATER SHELLS.

For these lessons, it has been deemed best to commence with the shells of mollusks, such as snail-shells and mussels. They are better objects to examine than insects, being more simple in structure, and less liable to be broken in handling. When found alive, their habits can be readily studied, as they can easily be kept alive in jars filled with water.

1. Let the pupils first make a collection along the shores of some lake or river, picking up all the different kinds of shells they meet with. The waves will have thrown them up on the shores, or in times of drought the waters will have left them exposed. Certain kinds are very small, though they will be found by sharp looking. Most of the shells collected will be empty, and these shells are called *dead shells*, because the soft-bodied creatures once contained in them have died and decayed, leaving the hard, limy shells. Some of the shells collected may contain the animal, and at one time all of them possessed a little creature within, which was the fabricator of the shell.

2. Remember that the shell is not a house built by the snail, as a wasp builds its nest, but the shell is a part of the animal, and is connected to it by certain muscles, so that it cannot leave the shell, as many suppose.

The empty, or dead, shells, are to be studied first.

Looking over the shells collected, we shall find some of the following kinds:

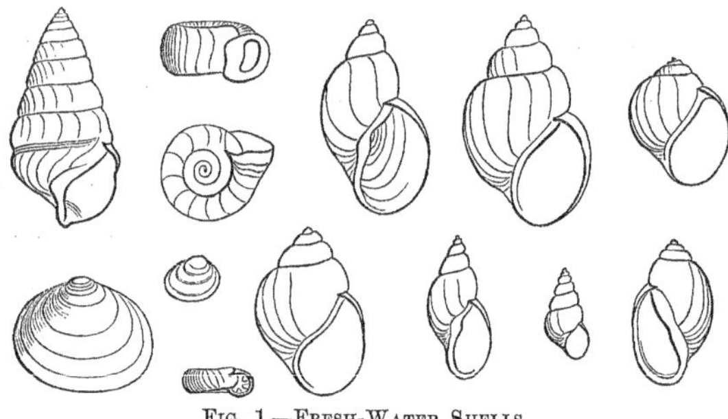

Fig. 1.—Fresh-Water Shells.

A number of fresh-water mussel-shells, also, will probably be collected.

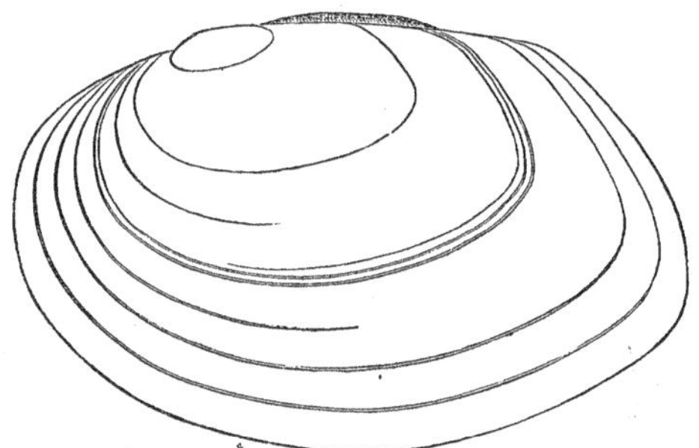

Fig. 2.—Fresh-Water Mussel-Shell.

These are to be reserved for future examination. Those

having a spiral turn or twist are called snail shells, and are to be studied first.

3. Let the pupils pick out from their collections the shells like these:

FIG. 3.—FRESH-WATER SNAIL-SHELLS.

The different spiral turns, or twists, are called *whorls*, and the whorls together form the *spire*. The opening into the shell is called the *aperture*, and the line separating the whorls is called the *suture*. The pointed end of the spire is called the *apex*.

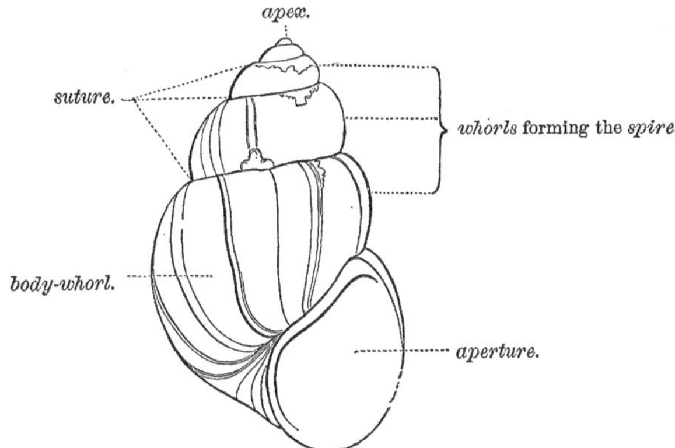

FIG. 4.—SHELL, WITH PARTS NAMED.

In some shells the spire is *elongated;* in others the spire is *short;* in others still the spire is *depressed* or *flattened.*

Fig. 5.

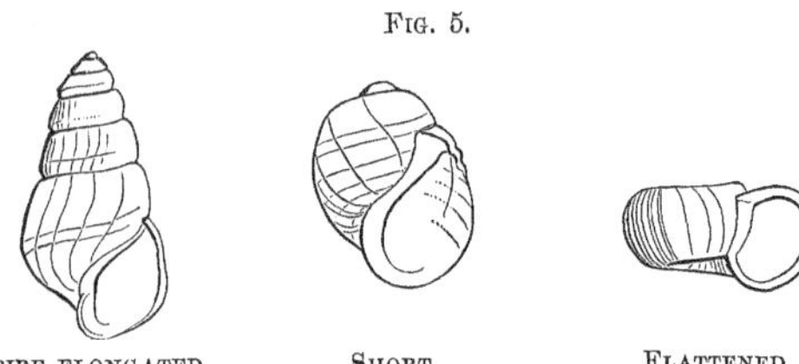

Spire elongated. Short. Flattened.

4. If the shell is held in the hand, with the aperture toward the holder, and the spire pointing upward, as in the figures drawn, the aperture will be either toward the right hand, or toward the left hand. In the figures already given, the aperture is on the right hand, and these shells are called *dextral*, or right-handed, shells.

Shells having the aperture on the left hand when held in the way above described, are called *sinistral*, or left-handed, shells. Let the pupils here examine all the shells they have collected, holding each one with the spire pointing upward, and the aperture toward them, and separate the *dextral* shells from the *sinistral* shells. As sinistral shells are not as common as the other kind, it may be that none will be found in the first collections made by the pupils. The following is a figure of a sinistral shell:

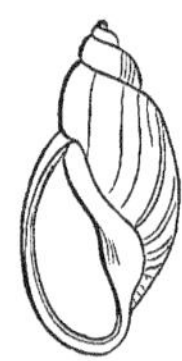

Fig. 6.—Sinistral Shell.

5. If the surface of the shell be examined closely, delicate lines running from one suture to another will be seen, as in the figures already given; and, if the shell be looked at from the side of the aperture, these lines will be found running parallel to the edge of the aperture, or *lip*, as it is called.

These delicate lines are called *lines of growth.*

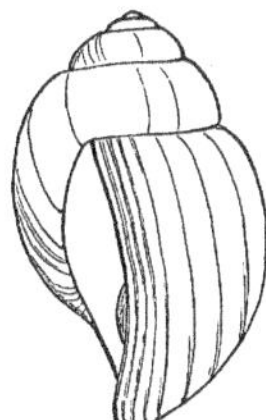

FIG. 7.—SHOWING LINES OF GROWTH RUNNING PARALLEL TO THE EDGE OF THE APERTURE.

The shell is increased in size by successive layers of shelly matter added to the borders of the aperture. In this way the shell grows.

A clearer idea of the growth of a shell may be obtained by studying the next figure (Fig. 8): *A* representing in outline a young shell; *B* representing the full-grown shell in outline; and *C* representing the same outline as *B*, with a number of lines of growth represented upon it.

If the shell were now to continue its growth a single half-turn, or *whorl*, the dotted lines would indicate the increased stages it would assume: *a* representing the first increase in size, *b* the next stage, and *c* the appearance of the shell when the additional half-whorl has been completed.

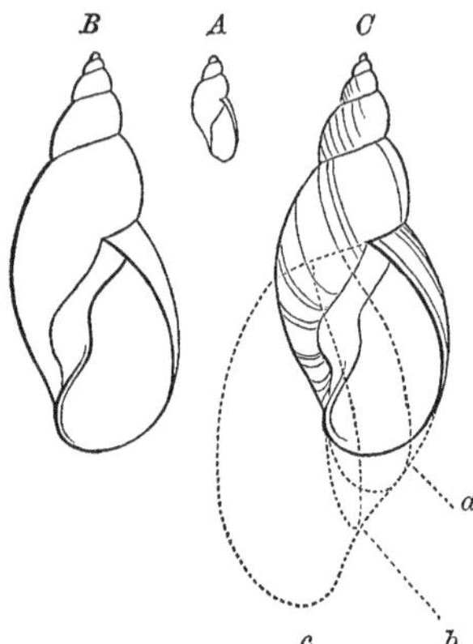

FIG. 8.—ILLUSTRATING THE INCREASE IN SIZE OF A SHELL.

Among the lot of shells collected by the pupil, different-sized ones will be found. Now, if a number of these, of different sizes, can be picked out, provided they belong to the same kind, or species, it will be noticed that the apex of all of them will be the same; but that the shells have increased in size at the aperture, and the aperture will be larger, and the larger shells will have more whorls than the smaller shells. The following figures illustrate four different ages of the same species of shell:

FIG. 9.—ILLUSTRATING DIFFERENT AGES OF THE SAME SHELL—THE LOWER FIGURES REPRESENTING A VIEW OF THE SHELL FROM THE APEX.

6. The axis around which the whorls revolve is called the *columella*. This axis is generally solid, though in many shells it is hollow, as if the whorls had turned around a shaft which had afterward been withdrawn. This hollow axis looks like an opening in the base of the shell, as in the following figure:

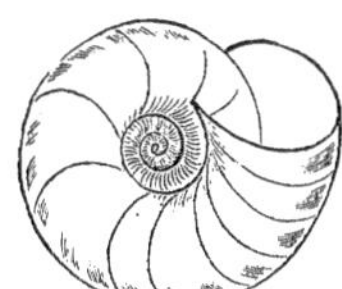

FIG. 10.—A SNAIL-SHELL SEEN FROM BELOW.

This opening is called the *umbilicus*. The apex of the shell is sometimes called the *nucleus*, because the shell commences to grow from this point.

CHAPTER II.

FRESH-WATER SNAILS.

7. THE pupils will now be required to bring in some live snails. Let them examine bits of bark, chips, or branches, found in ditches, or muddy brooks. Under lily-pads and on the stems and leaves of other aquatic plants, and on stones in rivers, snails of various kinds will be found. A dipper with the bottom perforated, or made into a sieve, and attached to a wooden handle four or five feet in length, will be found useful in scooping up the sand or mud from the

bottom of rivers and ditches. The dirt having been sifted out, the shells and other objects will be left behind. The dipper may be made as in the figure.

FIG. 11.—DIPPER ATTACHED TO A WOODEN HANDLE FOR COLLECTING SNAILS.

Shells collected with the snails inside, and cleaned for the cabinet, are called *live shells.* They are always more fresh and perfect than dead shells.

Having made the collection, the snails should be kept alive in a wide-mouthed jar, or bottle, care being taken not to have more than fifteen or twenty in a jar holding a quart of water.

8. The pupils will have secured some of the following forms:

FIG. 12.—FRESH-WATER SNAILS.

The broad, creeping disk upon which the snail rests, and by which it retains its hold to the glass, is called the *foot.* The snail moves about, and crawls or glides slowly along, by means of the foot.

The two little horns or feelers, in front, are called *tentacles*, and, as the snail moves, the tentacles are seen stretched out in front, and occasionally bending, as if the creature were feeling its way along. The eyes are seen at the base of the tentacles, as two minute black dots. The mouth is between the tentacles, and below. The part from which the tentacles spring is called the *head*, and the opposite end of the body is called the *tail.* The surface upon which the snail rests is called the *ventral* or lower surface, and consequently that portion of the body which is above is called the *dorsal* surface, or back.

9. The pupil, in watching the habits of the snails he has collected, will notice some of them crawling to the surface of the water to breathe air. The snail accomplishes this by raising the outer edge of the aperture to the water's edge, and then opening a little orifice in the side, through which the air enters to the simple lung within.

This orifice is on the right side in those snails having dextral shells, and on the left side in those snails having sinistral shells.

Many kinds of snails which live in fresh water are called air-breathers, because they are forced to come to the surface of the water to breathe air. In doing so they first expel a bubble of air, which may be seen escaping from the breathing-orifice, as in Fig. 14, *B*.

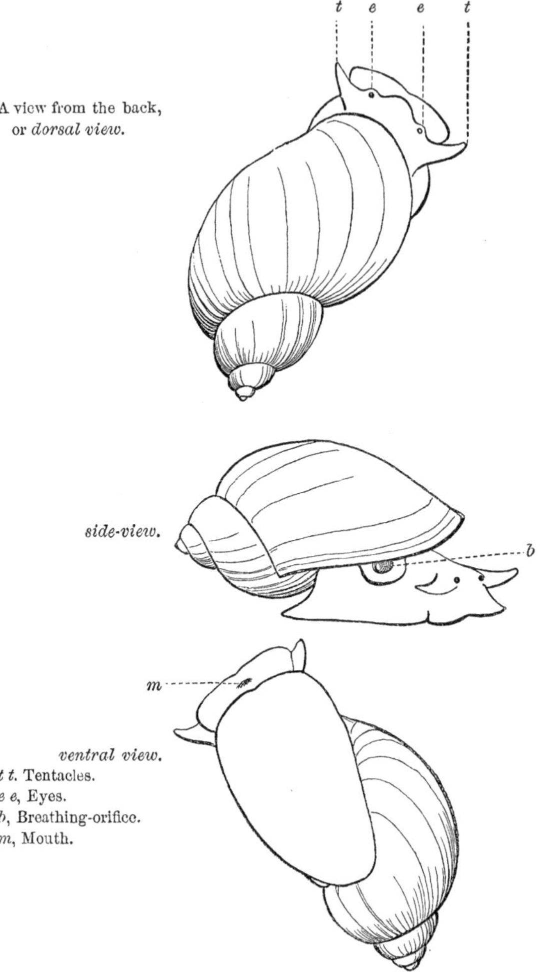

FIG. 13.—A FRESH-WATER SNAIL SEEN FROM ABOVE, FROM THE SIDE, AND FROM BELOW.

These fresh-water air-breathing snails may be kept under water for many hours before life is extinct.

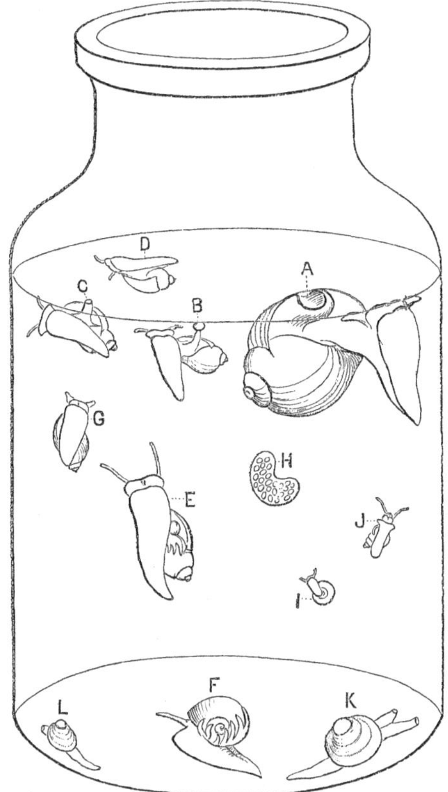

FIG. 14.—Jar of water, in which is contained a number of species of mollusks, some of which have already been studied. Some of them are near the surface, breathing air: *A* and *C* are taking in air; *B* is just expelling a bubble of air from the lung; *D* is crawling on the surface of the water; *E*, *G*, and *I*, are in the act of crawling up, to get a fresh supply of air; and *J* is a water-breather, having gills, but no lung.

10. Among the snails collected, there will probably be found some which have a peculiar scale on the hinder part of

the body. When the snail crawls, this scale will be seen just behind the shell, as in Fig. 15, *o*.

This scale is called the *operculum*, and when the snail has contracted, or drawn within the shell, the operculum is seen to fit the aperture of the shell, closing the shell as a stopper closes the mouth of a bottle.

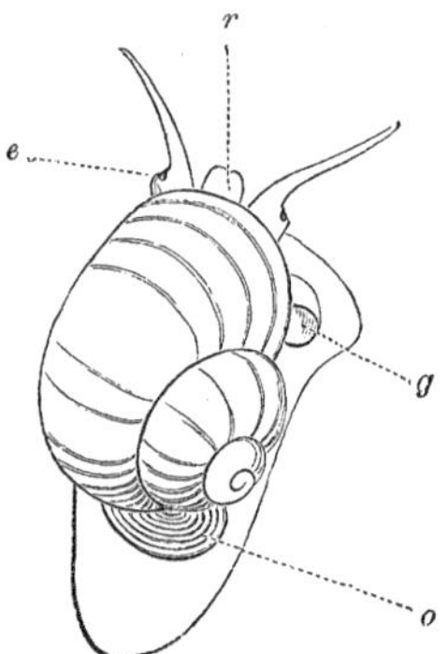

FIG. 15.—SNAIL WITH OPERCULUM.—*o*, Operculum; *e*, Eye; *r*, Rostrum; *g*, Entrance to Gill-Cavity.

Nearly all sea-snails, that is, snails which live in salt-water, and many species of fresh-water snails, and also many snails which live in damp places on the land, and which are called land snails, have an operculum.

When the snail has retired within the shell, the operculum will look like this in the aperture of the shell (Fig. 16):

A series of concentric lines will be seen marking the operculum, and these are the lines of growth, the operculum growing around the outer edge by successive additions, just as the shell grows by successive additions to its outer margin.

The Western rivers teem with species of snails having opercula.

11. If the pupil has any of these operculated snails alive,

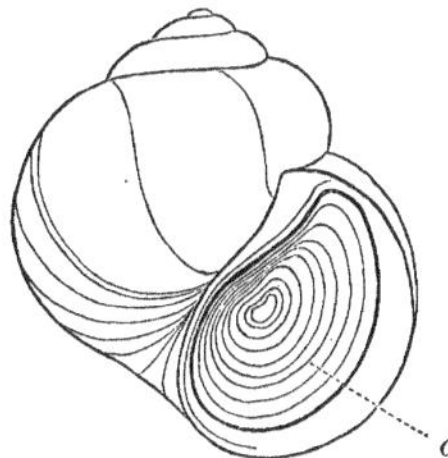

FIG. 16.—APERTURE OF SHELL CLOSED BY OPERCULUM, *o*.

he will observe that they do not come to the surface to breathe air.

Instead of a lung, they have a cavity containing an organ, or part, called the gill, by means of which the snail is capable of getting from the water what the air-breathing snail gets from the air, namely, *oxygen.*[1]

It will be seen that the head of the snail is shaped differently in the snails having an operculum, the mouth being at the end of a sort of proboscis or *rostrum*. (*See* Fig. 15.) The shells, too, are, as a general thing, more solid.

12. Thus far the pupils have examined those snails which live in fresh water. Some of these were air-breathers, and came to the surface of the water at intervals to breathe air. He has studied other fresh-water snails which did not breathe air directly, but performed this function by means of

[1] If the class is sufficiently advanced, the teacher may here explain about oxygen and what the blood requires, and gets by respiration.

an organ called the gill. And these snails were operculated, that is, they all possessed a little scale called the operculum, which closed the aperture tightly when the snail contracted within the shell.

He has also learned that the shells grow in size by successive additions of limy matter deposited around the free border of the aperture, and that the delicate lines which mark the surface of the shell, and which run parallel to the outer edge of the aperture, are lines of accretion, or lines of growth.

CHAPTER III.

LAND SNAILS.

13. THERE are many other species of snails which live out of the water altogether, though they are generally found in damp places; and these are called land snails, because they live on the land.

Let the pupils now endeavor to collect some land snails. By going to some hard-wood grove of maple, beech, or oak, and turning over the layers of dead leaves, old rotten logs, or pieces of bark, they will be sure to find some specimens of land snails. Some of them do not grow larger than a pin's-head, others have shells as large as a walnut.

They are generally light brown in color, and the smaller species often have highly-polished or shiny shells.

The spire is generally depressed or flattened. In many,

the border of the aperture has a thickened white rim, or lip, as it is called. Such a collection having been made, the pupil will find among them some of the following kinds:

FIG. 17.—THE SHELLS OF LAND SNAILS.

14. The snail, as it crawls along, leaves a slimy trail after it. This trail consists of a fluid, which not only flows from the creeping disk, but also from the surface of the body. If the back of the snail is irritated by a sharp-pointed stick, a little whitish mass of this fluid, or mucus, will adhere to the end of the stick.

By placing the snail on a piece of glass, and allowing it to adhere and crawl on it, a good view may be obtained of the peculiar movements of this creeping disk, by looking through the glass from the other side.

The breathing-orifice may be found just within the aperture of the shell, on the right side of the snail.

This orifice will be seen opening and closing at intervals. (*See* Fig. 18, *b*.)

15. In the fresh-water snails there are but two tentacles upon the head. In the land snails with few exceptions, the

tentacles are four in number, a larger and a smaller pair. The larger tentacles are called the *superior tentacles;* the

FIG. 18.—LAND-SNAILS CRAWLING.—*b*, Breathing Orifice; *s t*, Superior Tentacles; *i t*, Inferior Tentacles; *m*, Mouth.

smaller ones, often appearing as mere tubercles, are called the *inferior* tentacles. (*See* Fig. 18, *s t* and *i t.*) As the snail crawls, the superior tentacles are seen in constant motion, as if the creature were feeling its way about with them.

The eyes, instead of being at the base of the tentacles, as in the fresh-water snails, are found at the tip of the superior tentacles.

In the land snails, with few exceptions, the tentacles can be drawn within the head, and for this reason they are also called *retractile* tentacles.

While the snail is crawling, let the pupil touch the end of the tentacle with his finger, or, even if he alarm the snail by a sudden jar, he will see the tentacles quickly withdraw within the head. The pupil will observe that the bulbous end containing the eye disappears first, as the end of a glove-finger disappears, when the hand is withdrawn from the glove, the glove turning wrong-side out.

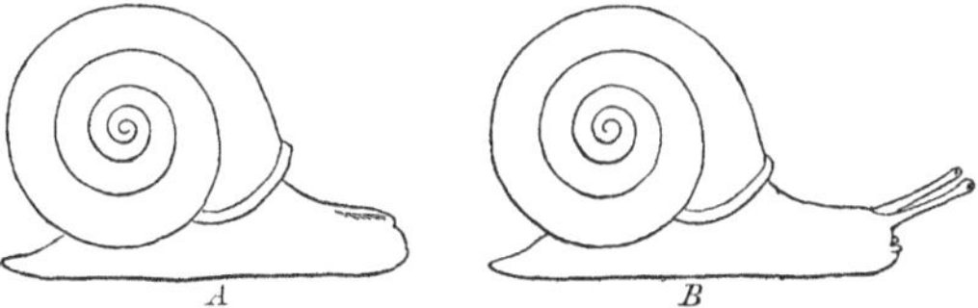

FIG. 19.—SHOWING SNAIL WITH TENTACLES RETRACTED, *A*; AND TENTACLES PROTRUDED, *B*.

16. Something may now be learned as to the way in which land snails eat.

By placing before the snail the tender leaves of lettuce or cabbage, the head will be seen to move, as little mouthfuls of the leaf are bitten off. The upper lip of the mouth is furnished with a hardened piece called the *buccal plate*. It is crescentic in shape, and, in some species, the cutting edge is notched, so that it acts like an upper set of teeth, by which it bites off little bits of the leaf. The floor of the mouth is lined by a membrane having upon it rows of little points which enable the snail to rasp and grind its food. These parts

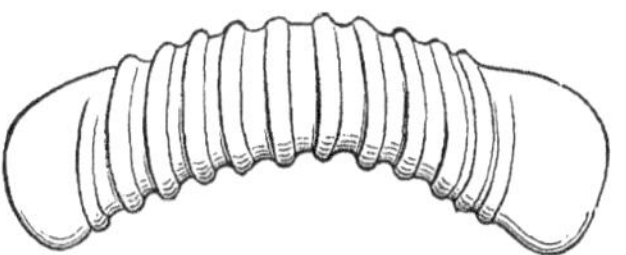

FIG. 20.—JAW, OR BUCCAL PLATE OF A LAND SNAIL, HIGHLY MAGNIFIED.—(It can just be discerned without a microscope.)

are so minute that they can be studied only by the aid of a microscope. If the pupil will watch his fresh-water snail as it crawls around the sides of the jar, he will see at intervals the mouth open, and a glistening tongue appear, as the snail laps up the scum which forms upon the surface of the glass.

NOTE FOR TEACHERS AND PUPILS.—Let the teacher here explain to the pupils what is meant by an object being magnified.

If the teacher has a common magnifying-glass, let each pupil in turn observe its magnifying effect, by looking at a common house-fly, or the printed page of a book. If a microscope can be shown the class, it will be better still.

Let it be explained, also, what is meant by an object being enlarged two, or three, or more times. To be enlarged two or three times, is to make the object two or three times as long as it was before, and of proportional bigness.

Oftentimes the object has to be reduced in size in the figure, as in pictures of large animals, in the picture of an elephant, for example.

In representations of very small animals, however, the figure has to be enlarged in order to show parts plainly that could not otherwise be seen, were they drawn natural size, that is, a size corresponding to the actual size of the animal. Thus in Fig. 24, *b* and *c* are greatly enlarged, to show the little snail within the egg.

17. In searching for snails, the pupil will come across snail-like animals, which have no coiled shells on their backs. Let the pupil examine the under side of damp boards or plank walks in gardens, and he will be sure to find them.

They are very common in old gardens in cities. These

FIG. 21.—A SLUG CONTRACTED.

creatures will be found clinging to the board or upon the ground, and will present this appearance (Fig. 21). Soon, however, they will stretch out their tentacles, and commence to crawl, and then their resemblance to the shell-bearing snails will be seen at once.

Instead of having a coiled shell into which they can retreat when alarmed, they have a little limy scale imbedded in a portion of the back, called the *mantle.* The breathing-orifice is on the right side of the body, and the tentacles, mouth-parts, creeping disk, and other features, are quite similar to the land snails already studied.

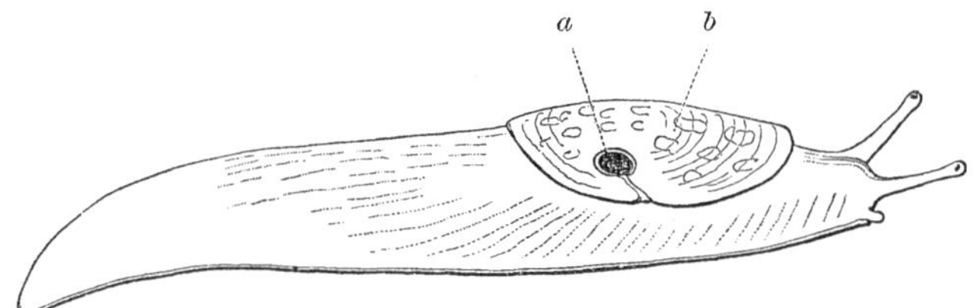

FIG. 22.—A NAKED LAND SNAIL, OR SLUG, FULLY EXPANDED.—*a*, Mantle; *b*, Breathing-Orifice.

18. On the approach of winter, land snails bury themselves in the ground, and those that have shells retire within the shell as far as possible, and close the aperture of the shell with a film of the mucus which the body secretes so abundantly. In this condition they remain dormant until the warm weather of spring revives them again.

If the pupil will collect a lot of snails in the early spring, and keep them confined in a box, with earth, damp leaves, or bits of rotten wood or bark, the snails in the course of a few weeks will lay a number of little eggs. These eggs will be white and round, about the size of a pin's-head. By careful tending, that is, by keeping the leaves slightly moist, the eggs will hatch out tiny snails, and these will attain half their mature size the first season.

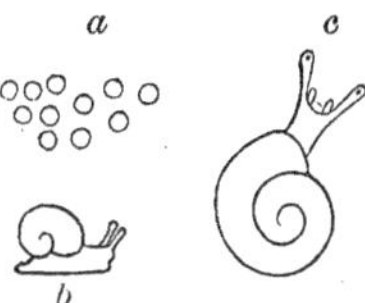

FIG. 23.—*a*, Eggs of Land Snail; *b*, Young Snail just hatched; *c*, Young Snail somewhat advanced: *b* and *c* are magnified.

19. If the pupil will also collect a lot of fresh-water air-breathing snails, and keep them alive, they will deposit their eggs upon the sides of the glass jar in which they are confined. These eggs will be oval in shape, and transparent, and will be inclosed in a transparent, jelly-like substance. Fig. 14, *H*, represents the appearance of a cluster of these eggs.

Fig. 24 shows a cluster of eggs with the appearance of two eggs highly magnified, showing the young snails as they appear within the egg.

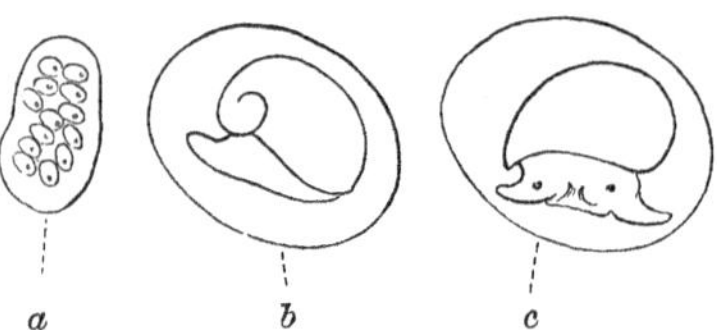

FIG. 24.—*a*, Cluster of Eggs of a Fresh-Water Snail; *b*, *c*, Eggs enlarged, showing the young Snails within the Eggs.

With the aid of a magnifying-glass, the eggs may be watched from day to day, and the young snail can be seen in its various stages of growth.

20. If a land-snail is taken out of its shell (and this can be done if boiling water be first poured upon it, and then with a pin the animal can be readily picked out), it will present this appearance (Fig. 25):

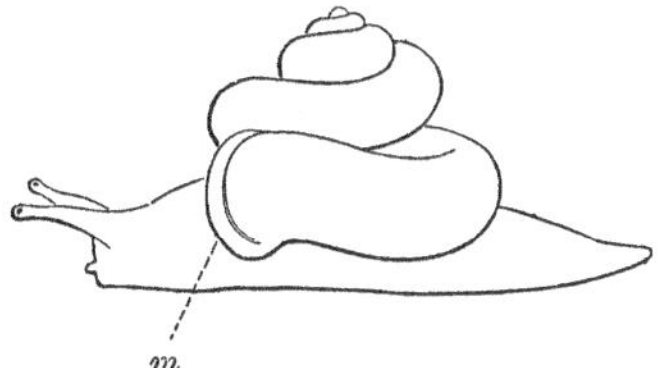

FIG. 25.—LAND SNAIL REMOVED FROM ITS SHELL.—*m*, Mantle.

The portion contained within the shell presents the same general appearance as the shell itself. A free border, or collar, is seen which corresponds to the aperture of the shell. This border is called the *mantle*, and is a characteristic feature of all the snails thus far studied. It is the edge of the mantle which deposits the successive layers of the shell, and increases its size. In the slug, there is only the limy scale; this is buried in the mantle, which is plainly seen covering a portion of the back, like a shield (Fig. 22, *a*).

CHAPTER IV.

SEA SNAILS.

21. CLASSES that live near the sea-coast may now study the marine, or salt-water snails. These may be collected alive at low tide, upon rocks, or under the sea-weed. By

examining pools left at low tide, many little sea snails may be seen creeping about.

A good place to collect dead shells may be found along an exposed beach. After a violent storm, when the waves have been running high, a great many animals are thrown up from the sea, and among them many shells may be secured.

The following will be some of the shells collected:

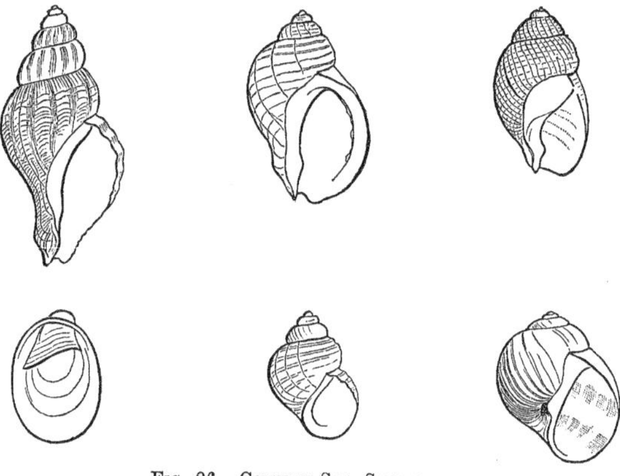

FIG. 26.—COMMON SEA SNAILS.

22. With very few exceptions, all sea snails are water-breathers; that is, they are furnished with gills, instead of a simple lung. Most of them are operculated.

In the shells collected the pupil will find two well-marked groups.

In the two lower right-hand figures of Fig. 26, the shells have an aperture with a continuous border; that is, there is no notch, or fold, in it. In the three upper figures there is a notch, or fold, in the base of the aperture.

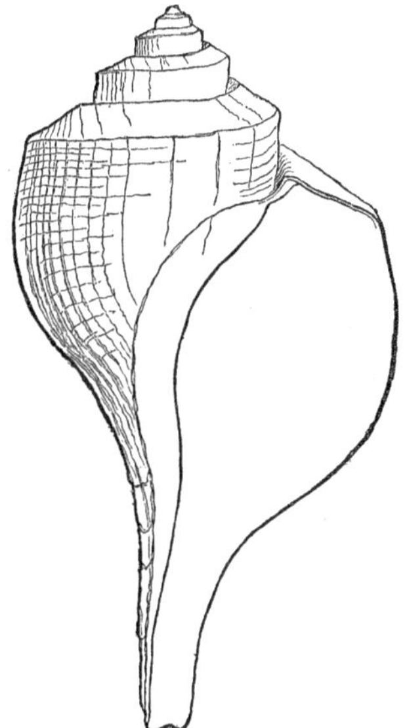

FIG. 27.—SHELL SHOWING LONG CANAL.

This notch is called the *canal*, and in some species it is very short, while in other species it is very long, as in Fig. 27.

The mantle of the animal is prolonged in a fold which occupies this canal, and is called the *siphon*. Through this fold or *siphon* the water finds access to the gills. (Fig. 28 shows another species. The siphon is seen as a fold of the mantle running into the canal of the shell.)

23. A very common species, found in the greatest abundance from Maine to Florida, on mud-flats, will give a good

illustration of the uses of the siphon. In this particular species, the siphon is much longer than the canal, and, when the snail is crawling, the siphon is bent upward. As the

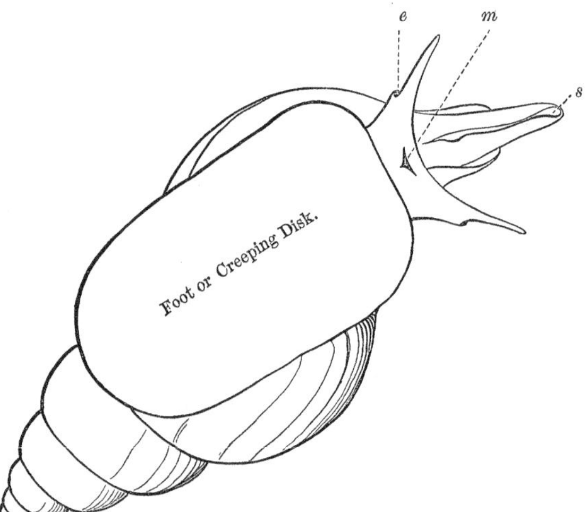

FIG. 28.—A SEA SNAIL SEEN FROM BELOW.—*e*, Eye; *m*, Mouth; *s*, Siphon.

habit of this species is to crawl along partly buried in the mud, the siphon, projecting above the level of the mud, conducts the pure sea-water to the gills of the snail below. Fig. 29 illustrates the appearance of this snail:

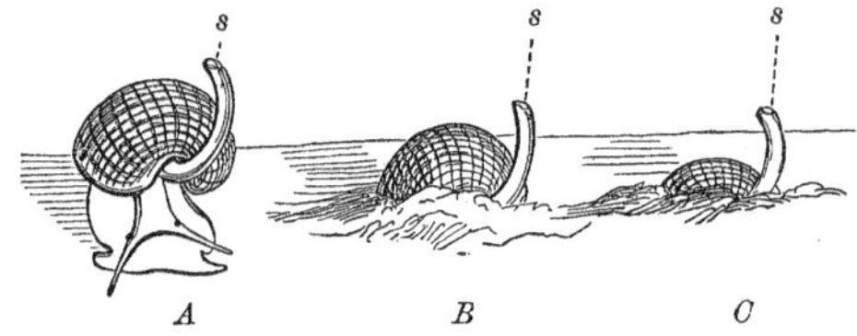

FIG. 29.—*A*, the Snail crawling upon the Surface of the Mud; *B*, the same slightly buried; *C*, the same nearly buried; the Siphon, *s*, is seen curved upward.

Such shells are called canaliculated shells.

The aperture of the shell is said to be *entire*, when it does not possess this notch, or canal. Let the pupil separate the shells having the aperture entire, from those shells having a canal.

These differences in the shell, as slight as they appear, are accompanied by corresponding differences in the character and habits of the animal.

Those snails having the aperture of the shell entire are with few exceptions vegetable-feeders, while those having a canal to the shell are flesh-feeders. The mouth-parts, and opercula, too, are different in the two groups.

24. Other shells will be found differing greatly in appearance from those thus far studied. One of these is represented in the lower left-hand corner of Fig. 26.

Another species, called the limpet, looks like this (Fig. 30):

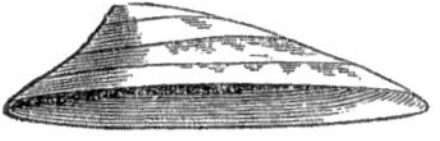

Fig. 30.—Shell of Limpet.

These shells will be found sticking with great tenacity to the rocks, and some skill and force will be required to remove them. This can be done by using the large blade of a pocket-knife, and suddenly scraping them from the rock. If they are then placed in a saucer of water, with the shell downward, the animal within will be found to have the broad, creeping disk, head, tentacles, and other parts, peculiar to the snails already studied.

25. In the land snails, it was learned that the eggs were deposited separately, while in the air-breathing, fresh-water snails the eggs were inclosed in a gelatinous substance. Among the sea snails there are many species which inclose their eggs in pods, or capsules, as is also the case in the operculated fresh-water snails. Sometimes the capsules are clustered together in large masses, as in those of the whelk (Fig. 31):

Fig. 31.— A very Small Cluster of Eggs of the Whelk deposited on the Stem of a Large Sea-Weed.

In others they are united in a long string (Fig. 32).

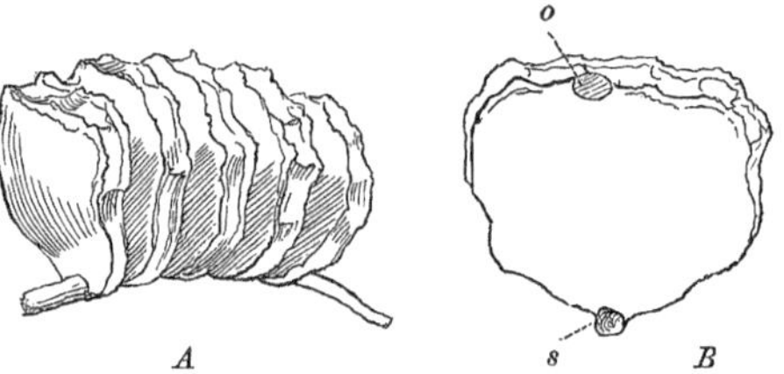

Fig. 32.—*A*, a Portion of a String of Egg-Capsules, from a Florida Species of Whelk similar to Fig. 27; *B*, a single Capsule separated, showing Outlet, *o*, through which the Young escape; *s*, Stem.

The common cockle sticks its separate egg-capsules side by side upon the rock (Fig. 33).

The beach-cockle deposits its eggs in a broad ribbon of sand cemented together, looking very much like a deep saucer, with the bottom broken out, and the side separated (Fig. 34).

FIG. 33.—EGG-CAPSULES OF COMMON COCKLE.—(A shell of the animal which produces these capsules is shown in the upper central figure of Fig. 26.)

This ribbon is elastic when wet, and, if it is held up to the light, the little transparent spaces for the eggs will be plainly seen.

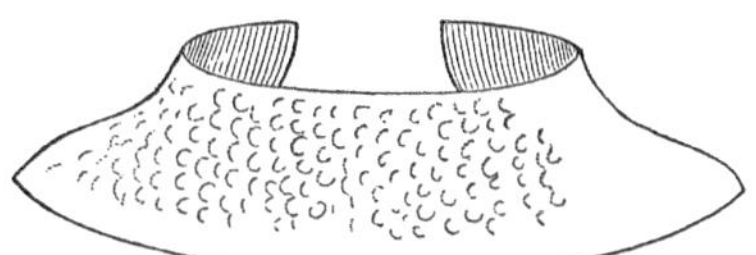

FIG. 34.—EGG-RIBBON OF BEACH-COCKLE (the shell of which is shown in the lower right-hand figure of Fig. 26).

CHAPTER V.

FRESH-WATER MUSSELS.

26. LOOKING over our fresh-water shells again, we find many that are known as mussels, or clams. These shells are common everywhere along the margins of brooks, rivers, and lakes. The musk-rats feed upon the soft parts of the mus-

sels, and the remains of their feasts may be found in piles of mussel-shells all along the shores of certain lakes.

The shell is composed of two pieces, or *valves*, as they are called. The two valves are often found united, and the margin along which they are connected is called the *hinge-margin*, because the shells hinge at this part, and will open and shut as a door swings upon its hinges.

Let the pupil now examine a perfect fresh-water mussel, that is, a mussel in which the valves are united in this way, and he will observe that they are connected by a brownish substance, which is quite elastic when the shell is alive, but becomes brittle when dried. The shells are held together as the covers of a book are held together by the back.

This substance is called the *ligament*, and the position of this ligament will indicate the back, or *dorsal region* of the animal.

27. On the outside of the shell will be seen fine lines, which run nearly parallel to the outside margin of the shell. These lines are the *lines of growth*, and indicate the successive stages of growth, or increase of the shell, as in the lines of growth in the snail-shell already studied, and, as in the snails, the growth takes place at the margin of the shells.

The pupil may trace these concentric lines back, as they grow smaller and smaller, till they are found to start from one point at the back of the shell, and this point is called the *beak* or *umbone*. It represents the starting-point in the growth of the shell. In fresh-water mussels, the umbones are eaten away by some corrosive action of the water, and the early stages in the growth of the shell are usually destroyed. In very young shells, however, the early stages can be plainly seen.

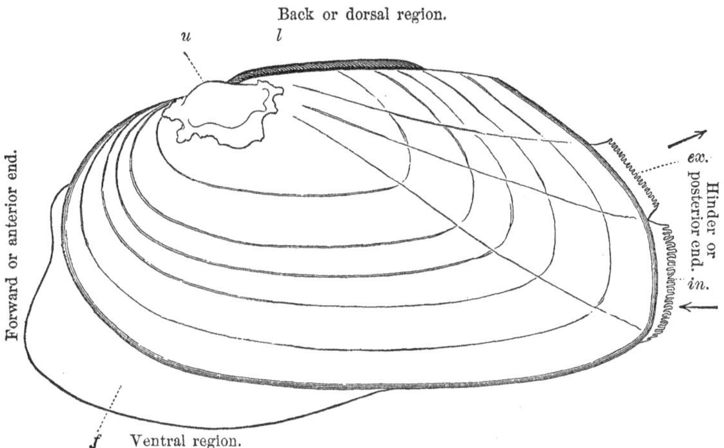

FIG. 35.—A FRESH-WATER MUSSEL.—*l*, Ligament; *u*, Umbone; *f*, Foot; *ex.*, Excurrent Orifice; *in.*, Incurrent Orifice.

28. The ligament is always behind the beak, or umbone, in fresh-water mussels, and in nearly all *bivalve* shells (so called, because they have two valves or pieces, while the snail-shells are sometimes called *univalve* shells, because they have but one valve or piece).

Let the pupil now hold a perfect mussel-shell in his hand (that is, a mussel in which both valves are together, and united across the back), with the ligament uppermost, and the umbone away from him, or beyond the ligament, and the valve on his left hand is the one which covers the left side of the animal, while the valve on his right hand covers the right side of the animal. The forward end will be the end away from him, and the hinder end will, of course, be the end toward him. (*See* Fig. 35.)

29. Let the pupils now endeavor to collect some fresh-water mussels alive. These may be found partly buried in the sand or mud of rivers and lakes. As they crawl along partly buried in this way, they plough up the sand, leaving a well-marked furrow or groove behind them. Every boy that goes in bathing is familiar with the peculiar furrow left by the fresh-water mussel. By following such a furrow, the mussel that made it will soon be found.

Fig. 36 represents the appearance of a common fresh-water mussel in the act of crawling.

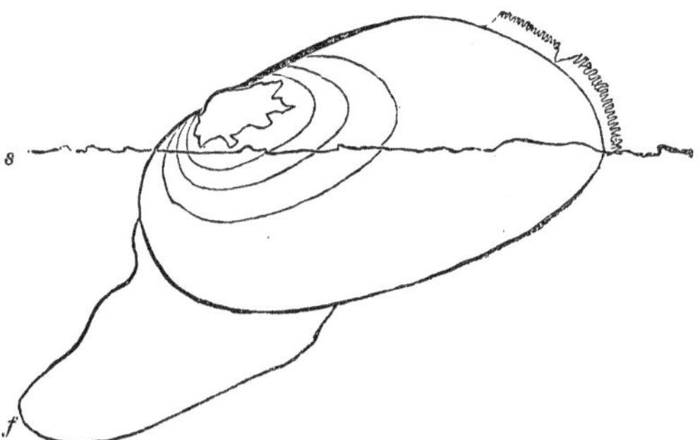

FIG. 36.—SHOWING POSITION OF MUSSEL WHEN CRAWLING.—*f*, foot buried below the surface of the sand *s*. Above the line *s* is supposed to be water, the line representing the bottom of a lake or river.

Having collected a few in this way, they may be placed in a large, shallow pan of water, and allowed to remain quiet for a while. Gradually the shells will open a little, and from the hinder end a curious fringed border appears; on examining this border, it will soon be found that the border forms two openings which lead into the shell.

Great care must be taken not to jar the dish, or the table upon which it rests. The slightest jar will cause the shells to instantly close. If some indigo, or small particles of dirt, be dropped near these openings, currents of water will be revealed; one current pouring out of the opening nearest the back, and another current as steadily pouring in at the other opening. The opening into which the current of water is passing is called the *incurrent orifice*, while the orifice from which a current of water is passing is called the *excurrent orifice*. The incurrent orifice is sometimes called the respiratory orifice, because the water is taken in to supply the gills which are the breathing or respiratory organs of the mussel, and this orifice corresponds to the siphon in the sea snails already studied. This current of water, besides bathing the gills, also carries in minute particles which are floating in the water, and these particles are conducted to the mouth of the creature, and swallowed as food. At the opposite end of the shell from these openings, or the forward end, a whitish, fleshy mass will be seen protruding. This is called the *foot*, and corresponds to the foot or creeping disk in the snails. By means of this foot the mussel crawls through the sand.

The mouth is above the foot, and always concealed within the shell. In Fig. 35 the foot is shown, and also the excurrent and incurrent orifices, with arrows drawn to indicate the direction of the currents of water.

In some small species of fresh-water bivalves, the excurrent and incurrent orifices are prolonged into tubes, and then they are called *siphons*. Fig. 37 represents a common species which the pupils may find in muddy brooks and ditches. By using the long-handled dipper already described, some specimens will probably be found. They are quite small, from the size of a pea to that of a nickel cent. The siphonal tubes are prominent, and the foot is long and tongue-shaped, and the animal is very active in crawling about; also in Fig. 14 *K* and *L* represent two of these small animals with bivalve shells.

Fig. 37.

30. The foot of these creatures resembles in appearance and action the foot of a fresh-water snail, only there is no mouth nor tentacles in sight. These parts are present, but are never protruded beyond the edges of the shell.

When the fresh-water mussels are partly open, a fleshy border will be seen just within the edges of the shell, and this is the border of the *mantle*, and corresponds to the same parts described in the snails; the fringed membrane which formed the openings at the hinder part of the mussel is simply a continuation of the mantle.

When the shells are removed from the animal, the mantle will be found lining the shells, just as the blank pages line the inside of a book-cover. While the edge of the mantle deposits the successive layers, which increase the size of the shell, the entire surface of the mantle deposits the pearly substance which lines the inner surface of the shells, and which is so characteristic of the fresh-water mussels.

31. Grains of sand, or other particles, getting in between the mantle and the shell, are soon covered by layers of pearly substance poured out, or secreted by the mantle. In this way pearls are formed.

If pearls are broken open, a centre, or nucleus, will be found, consisting of some particle of dirt or sand, or some substance which had found its way by accident between the mantle and the shell, and around which the pearly matter has been formed in successive layers.

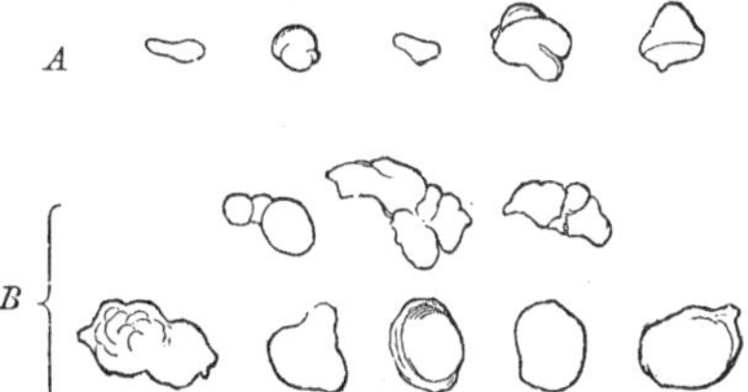

FIG. 38.—*A*, Pearly Concretions from a Fresh-water Mussel; *B*, Pearly Concretions from the Common Oyster.

In shells having a brilliant, pearly lining, or *nacre*, the pearls obtained are oftentimes very beautiful, and from certain Oriental species living in the sea, called *Avicula*, the most brilliant pearls of commerce are obtained. If, on the other hand, the nacre lining the shell is dull white, as in the common oyster, the pearls are dull-colored. This kind of pearls is often found in oysters.

The Chinese have long been familiar with the art of

making artificial pearls. By partly opening the shells of certain fresh-water mussels, and inserting little lead images, or other objects, between the mantle and the shell, the objects soon become covered with a natural layer of pearl.

32. Let the pupils now study the markings on the inner surface of the shells of river-mussels. The shells of these creatures are called *valves*, and are spoken of as right or left valves, according to whether they are on the right or left side of the animal.

Certain ridges and prominences will be seen at the hinge, and, when the valves are carefully joined, the ridges in one valve will correspond to grooves in the other valve. These ridges are called *teeth*. The short ones, near the beak, are called *cardinal teeth*, and the long ones *lateral teeth*. The margin upon which they occur is called the *hinge-margin*, for it is upon this margin that the valves turn. (*See* Fig. 39.)

33. Certain scars, or impressions, will be found marking the inside of the valves, and these indicate the point of the attachment of certain muscles to move the valves, and to enable the animal to protrude its foot, and crawl along.

These marks are hence called muscular marks, or *muscular impressions*, and will be found to correspond in the right and left valves.

An irregular, round impression will be found at each end of the valve, near the hinge-margin. These show where the muscles are attached to move and close the valves, and hold them firmly together. The muscles run directly across from one valve to the other; and, to open a live mussel, it is necessary to pass a sharp blade between the valves, and cut through the muscles, before the valves will open. These muscles are called the *adductor muscles*, and the scars or impressions on the valves are called the *adductor muscular impressions.* Very close to the adductor muscular impressions are seen smaller impressions, and these indicate where the muscles are attached which move the foot. These muscles are called the *pedal* muscles, and the impressions are called the *pedal muscular impressions.* One occurs just behind the anterior adductor impression; the other will be found just above, and in front of the posterior adductor impression.

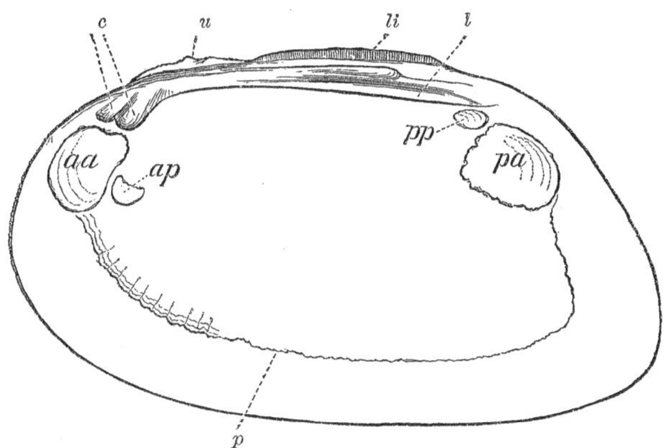

FIG. 39.—THE RIGHT VALVE OF A FRESH-WATER MUSSEL.—*c*, Cardinal Teeth; *l*, Lateral Tooth; *li*, Ligament; *aa*, Anterior Adductor Impression; *pa*, Posterior Adductor Impression; *ap*, Anterior Pedal Muscular Impression; *pp*, Posterior Pedal Muscular Impression; *p*, Pallial Line.

34. Besides these marks, the pupil will see a delicate and slightly irregular line running from the anterior to the

posterior muscular impression, just inside, and nearly parallel with the lower margin of the shell. This line is called the *pallial line*, and indicates where the mantle is attached to the shell. It will be observed that, when the soft parts are removed from the shell, the mantle adheres along this line.

The pupil may mark with a pen the names of all the parts upon the inside of a fresh-water mussel.

35. When the mussel is opened by separating the adductor muscles with a knife, the valves slowly open, and after the animal is removed the valves still remain partly open, and, to preserve them closed, a string has to be tied around them, and in this condition, if the ligament is allowed to dry, the valves will then remain closed. From this it is evident that the ligament acts upon the valves to draw them apart. To keep them closed, then, the animal must continually exert itself by contracting the adductor muscles; and it will be found that, when these creatures are left in the water, undisturbed for a while, the muscles relax, and the valves partly open. The ligament is elastic, and is stretched as it were from one valve to the other, over the back. A possible imitation of the action might be represented by partly opening the lids of a book, and then gluing across the back, from one lid to the other, a sheet of elastic rubber. If, now, the lids are tightly closed, the rubber is drawn out, or stretched across the back, and, if allowed to regain its elasticity, the lids are pulled apart. This experiment illustrates the way in which the ligament acts in those shells which have the ligament external.

CHAPTER VI.

CLAMS, MUSSELS, AND OYSTERS.

36. CLASSES having access to salt-water may now collect a lot of bivalves, as the clam, mussel, razor-shell, oyster, scallop, and whatever species they can find belonging to this group. A much greater variety of forms will be found in salt-water than in fresh-water.

Among some of the common species met with will be the following:

FIG. 40.—SALT-WATER BIVALVES.

In these the pupil may trace out the muscular impressions within the shell, and make out their relations to the impressions already described in the fresh-water mussels.

Many differences will be observed in the muscular impressions, as well as in the teeth and the position of the ligament.

37. As the common soft-shelled clam can be readily procured in the fish-markets, it will be well to study this first. A live specimen must be selected, and, as the clam lives a long time after it has been removed from the water, there will be no difficulty in getting the proper specimen. Upon pressing the valves together, or touching the soft parts which partly protrude from between the valves, the creature will show signs of life, by drawing the shells closer together, and this will assure the pupil that the specimen is alive.

A large shallow dish may now be filled with pure sea-water, and in this the clam may be placed. After it has remained there for some time, the black end of the animal, which is incorrectly called the head, will slowly stretch out from between the shells, and the end, unfolding, will display two openings fringed with little feelers (*see* Fig. 42). Into one of these openings the water will be seen flowing, while from the other a current of water will be seen issuing. And these openings are called the *incurrent* and *excurrent orifices*, and correspond to similar parts previously described in the fresh-water mussels. In the latter creature, the openings just protruded beyond the edge of the shell. In some very small species of fresh-water bivalves, one of which was shown in Fig. 37, these openings were at the end of separate tubes. In the clam the tubes are inclosed in one sheath.

The clam can protrude this apparatus to a length equaling that of the shell two or three times. As the clam lives buried at some depth below the level of the sand or mud in which it occurs, it requires this extension of the openings to reach the sea-water above.

38. It may be stated here, that the current of water passing into the general cavity of the shell not only carries the particle of food upon which the animal subsists, but conveys the pure sea-water to the gills by which it breathes, the gills performing the same function for animals living immersed in water as the lungs perform for creatures which breathe air. All bivalves depend upon currents of water to convey their food to them.

While, in the snails, the creatures could go in quest of food, having the power of protruding the head from the shell, and their mouths furnished with means to bite or rasp their food, in the bivalves there is really no head, they having only a little opening directly under the anterior adductor muscle, which is the mouth, and into which the particles of food are swept.

39. If, now, the clam is opened, the edges of the mantle will be found much thickened and united, except a small slit near the front edge, through which can be protruded a small, tongue-shaped foot. Powerful muscles will be found at the base of the united siphons or tubes, which move the siphons in and out, and an examination of the inside of the shell will show where these muscles are attached. The pallial line, instead of running directly from the anterior adductor impression to the posterior one, is abruptly curved back, and forms a sharp bend, as it turns again to the posterior ad-

ductor impression. This mark is called the *sinus*, and, when present, indicates the siphons to be of considerable size, and having large muscles to contract them.

Let these parts now be marked with a pen upon the shell, with their names, as in Fig. 41:

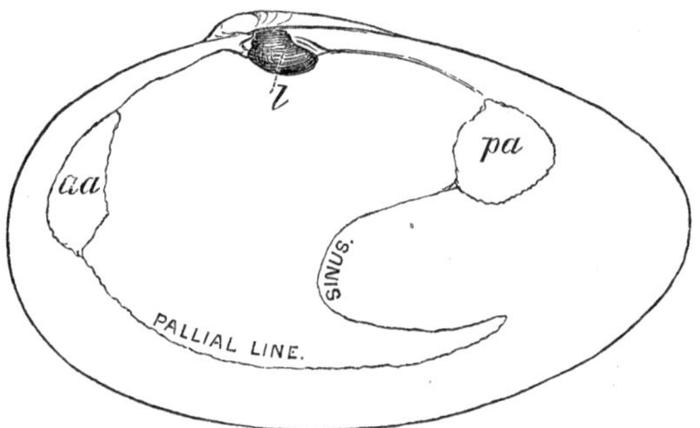

FIG. 41.—RIGHT VALVE OF A COMMON CLAM.—*l*, Ligament; *aa*, Anterior Adductor Impression; *pa*, Posterior Adductor Impression.

40. On opening the shell, a prominent tooth is seen on one of the valves near the hinge, while on the other valve there is a corresponding depression. When the valves are forcibly separated, there is left attached either to the tooth, or in the depression, a substance resembling dark glue, very elastic, and firmly attached to its place. This is the ligament, and is said to be *internal*, because it is within the shell, and not upon the outside, as in the fresh-water mussel already studied. When the animal closes the shell by contracting the adductors, the ligament is compressed, by being jammed between the prominent tooth, and the recess into which it fits. When the muscles relax, the ligament expands,

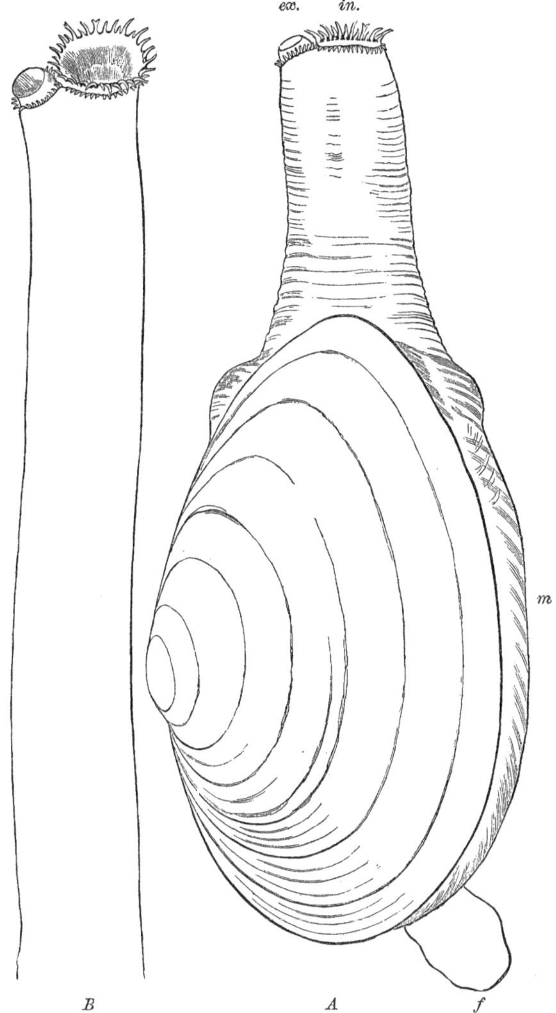

FIG. 42.—COMMON CLAM.—*A*, showing Siphons partly extended; *in.*, Incurrent Orifice; *ex.*, Excurrent Orifice; *f*, Foot; *m*, thickened border of the Mantle projecting beyond the Edge of the Shell; *B*, Siphons, greatly extended. (The shell is not drawn, as there was no room for it on the page.) The excurrent and incurrent orifices are more open than in *A*.

forcing the valves apart. The way in which it works might be illustrated by placing a piece of rubber inside the hinge of a door: when the door is closed, the rubber is squeezed, and the tendency would, of course, be for the rubber, in expanding, to again push the door open.

That it requires a continual effort for the clam to keep the valves closed, is seen in the fact that when these creatures are allowed to remain out of water for a while, as they are when in the market, the muscles get tired, and, relaxing, the shells partly open. If, now, the basket or barrel which they are in be suddenly shaken, the clams will as suddenly close, and a rustling sound is made, as the water is forced out from the gill-cavity, the water often squirting out in a stream from the siphonal openings.

41. On the rocks between high and low water mark, and adhering to the piles of the wharves, may be found clusters of mussels which are attached to these places, and to each other, by little brown threads which issue from between the valves below. These threads are made at will by the creature, one by one, and are fastened to the substances upon which they rest. The threads are called *byssal-threads*, and, combined, form the *byssus*.

If the pupils will collect a number of these salt-water mussels, and place them in a large glass dish or bowl filled with salt-water, they may watch the mussels as the creatures attach themselves to the sides of the vessel. In the figure, the byssus is seen like threads coming from between the valves below, with their ends adhering to the stone.

42. Fig. 44 represents an animal which is often thrown up on beaches along the coasts, after a storm, and whose shells are very common in the *débris* thrown up by the waves.

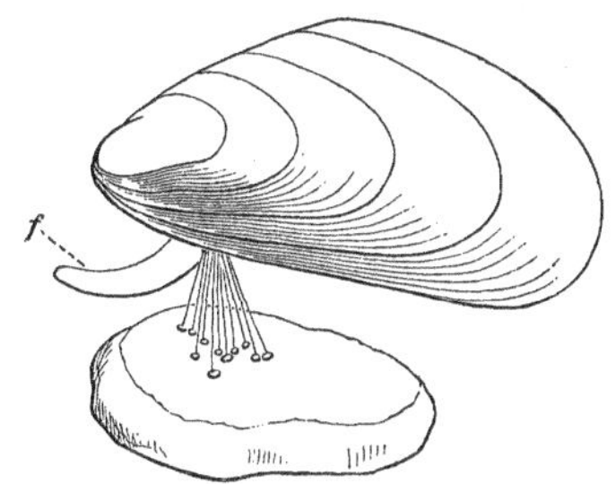

FIG. 43.—A MUSSEL ATTACHED TO A STONE BY ITS BYSSUS.—*f*, Foot.

These shells are very thin and delicate, and the valves are strengthened by a thickened rib which runs from the umbone toward the lower margin of the valve within.

In this species the siphons are united, the mantle projects beyond the edge of the valves, and the foot is flattened in front.

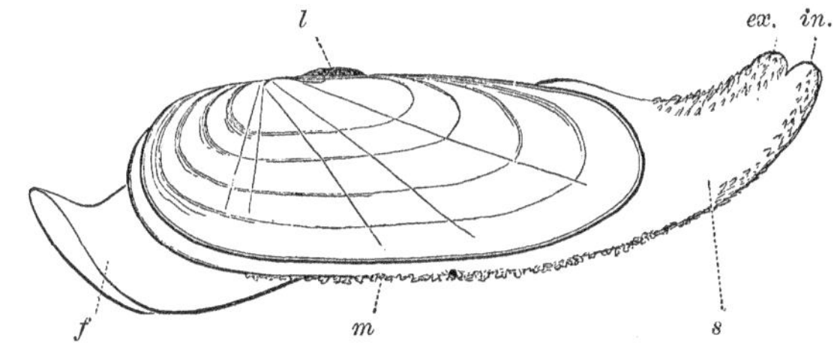

FIG. 44.—*l*, Ligament; *f*, Foot; *m*, Mantle; *s*, Siphons; *ex.*, Excurrent Orifice; *in.*, Incurrent Orifice.

Fig. 45 represents another animal which is often abundant on the sea-beach. In this species the siphonal tubes are separate, instead of being united. This figure represents the tubes only partly extended. They can be thrown out to twice the length represented in the figure.

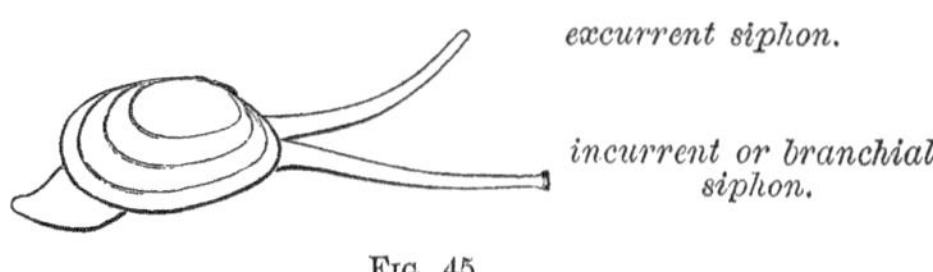

FIG. 45.

Another species quite similar to the above occurs on mud flats, in company with the common clam. If this be collected alive and placed in sea-water, the creature will extend its siphons, which are long and separate, and bend them in coils.

43. The pupils have now learned, among other things, a few features regarding the position which certain bivalves occupy in their native haunts: the fresh-water mussel creeping by means of its foot through the mud or sand in which it lives partly buried; the salt-water mussel, fastened to some place by means of its byssus; the soft-shelled clam, lying buried at some depth in the mud, and extending its siphons to conduct the pure sea-water to its gills, and food to its mouth.

The oyster differs considerably from these animals to which it is related, and which have just been studied. Instead of being free, they grow attached by one of their valves to the rock and to each other; clusters of a dozen or more individuals of different sizes are found growing, attached to each other, and forming large masses. At any oyster-market the pupil may get these clusters. Before studying a specimen, it is best to clean the shell thoroughly in water by means of a coarse brush.

44. Instead of having two adductor muscles, it has but one (though this muscle, it seems, is composed of two elements). A single dark-purple mark on the inside of each valve shows the point of attachment of the adductor muscle. When the oyster is opened, the mantle contracts somewhat, so that the edge of the mantle is some way from the margin of the shell, as shown in Fig. 47.

The left valve is the larger, and is the one that becomes attached; the right valve is flattened, and somewhat smaller. The mantle has its margins free; that is, the edges are not united as in the common clam, where they are not only united, but greatly thickened. Neither is the mantle prolonged into siphons as in other species; consequently, the water flows in at one portion of the shell, and pours out of another portion, not being definitely conducted by special channels, as in those forms heretofore given. The oyster can be readily studied, as specimens may be got in almost every village in the country.

In looking over canned specimens, be sure and pick out a large one, and one that does not appear to be mutilated, as they frequently are when taken out of the shell by the oysterman, or jammed, as they often are in packing.

To those who can get them alive, it is well to have the oysterman open the specimen, being sure that he removes the larger valve, leaving the oyster attached, and resting in the smaller and flat valve, which is the *right* one. To examine it properly, the specimen must be placed in a deep saucer filled with water, so as to cover it. A number of rinsings will remove the mucus with which the oyster is covered, and this will render the specimen in better condition to examine. In placing it under water in this way, the membranes float apart, and can be more readily studied.

45. The adductor muscle is near the middle of the animal. It is composed of two elements, one half being a glistening white, and the other half being grayish. Immediately adjoining the grayish portion of the muscle, a translucent space is seen, and this space contains the heart, composed of a body constricted in the centre, as if a tube had been tied in the middle by a string. This is the heart proper, and in specimens freshly opened the heart may be seen to slowly pulsate, or beat.

By raising the mantle, the gills will be seen as delicate, leaf-like membranes.

At the smaller end of the oyster, and that portion which comes next to the beak or hinge, the mouth will be found having on each side two delicate lappets, which are called the *palpi*. It will be difficult to find the mouth, and some patience will be demanded in lifting the mantle and following up between the palpi to where the mouth is.

The dark region just back of the mouth contains the stomach and liver; the dark or blackish portion, showing so conspicuously in cooked specimens, being the liver.

By referring to the accompanying figures, these parts may be readily made out :

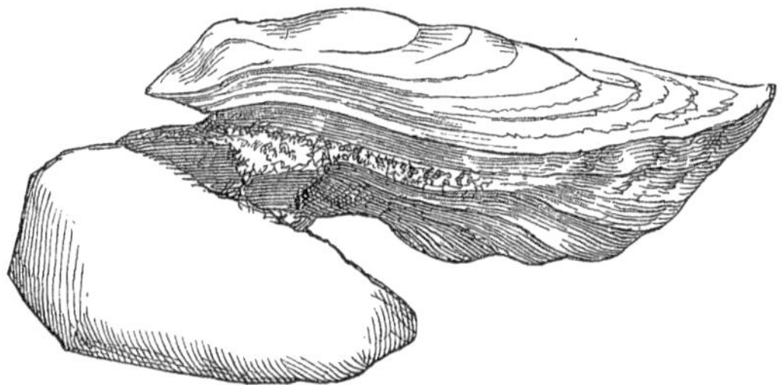

FIG. 46.—OYSTER ATTACHED BY LEFT VALVE TO A STONE.

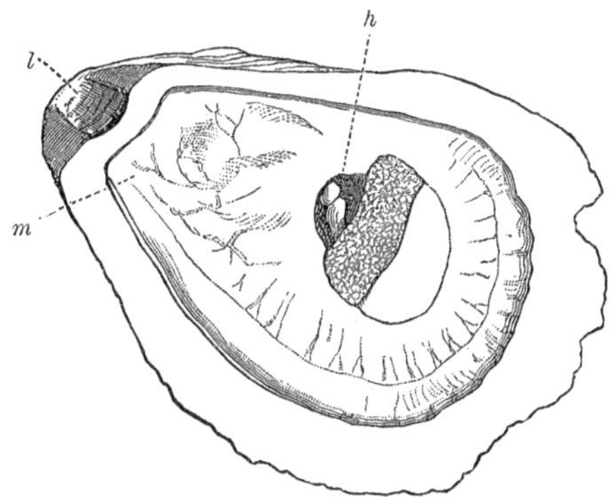

FIG. 47.—OYSTER WITH THE LEFT VALVE REMOVED.—*h*, Heart; *l*, Ligament; *m*, Position of Mouth.

46. The pupil will now recall some characters in common between the snails, and the mussels, clams, and oysters, thus far studied in these lessons, namely : they all have the body protected by a limy shell (except the slug), this shell either composed of one piece, as in the snails, or of two pieces or

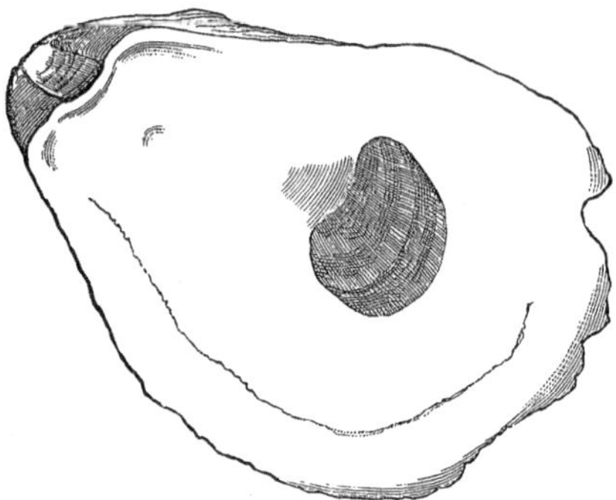

FIG. 48.—RIGHT VALVE OF AN OYSTER, the Dark Semicircular Mark near the Middle of the Shell being the Adductor Muscular Impression, the Pallial Line showing faintly.

valves, as in the clams, mussels, and oysters. All of them increase the size of their shells, or grow, by the addition of layers of shell-material to the edge of the aperture, or the margins of their valves, and these layers are indicated by delicate lines seen on the outside of the shell, and called lines of growth. They all have the creeping disk, or foot (excepting the oyster). In the snails, this is broad and flat; in the mussel and clam the foot is flattened sideways, and variously shaped. In the snails, the creature projects, with the foot, a head furnished with feelers, or tentacles, and the mouth is possessed of certain hard parts by which food can be eaten. In the mussels and clams there is no definite head, the mouth being hidden away within the mantle, and the creature projecting, from the forward end, only the foot. In all of these animals thus far studied there is a cavity within, containing the gills to which water has access, or else there is a simple lung, as in the air-breathing snails. These, with the cuttle-fishes, which we will not consider here, belong to a branch of the Animal Kingdom called Mollusca.

CHAPTER VII.

COLLECTING INSECTS.

47. These lessons, as well as the preceding ones, are prepared with the understanding that the pupils shall make a collection as far as possible of the species of animals studied. In fact, it is a part of the lesson to know how and where to collect, and above all to know how to preserve the specimens collected. To enable the pupils to do this, the briefest directions are given for the making of boxes, nets, etc., accompanied with the simplest methods of preserving the collections made.

In many cases the directions given are by no means the professional methods; thus the pupils are directed to use common pins for insects, while the professional collector uses only the true insect-pins made expressly for the purpose, but these are oftentimes difficult to procure, and are more expensive than the common ones.

In commencing these lessons, each pupil must first be provided with a number of common pins, and a box properly arranged in which to pin the insects collected.

Some holiday afternoon, or an hour before school-time

in the morning, may be spent in making the insect-boxes. These may be of any convenient size, having a depth of not over two and a half or three inches, and furnished with a lid. A shallow cigar-box will answer the purpose. The bottom of the box may be lined with strips of corn-pith, or slices of cork, into which the pins can be easily stuck. Large cork-stoppers will do, and these may be cut into lozenge-shaped pieces like this:

FIG. 49.—SLICED CORK FOR INSECT-BOX.

These pieces are to be fastened to the bottom of the box by gluing. If strips of corn-pith are used, they may be tacked or glued to the bottom of the box. The box, when finished, will look something like this:

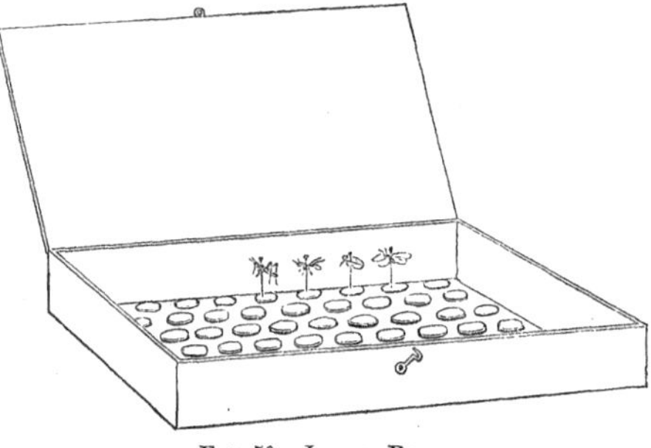

FIG. 50.—INSECT-BOX.

48. The insects, when collected, are to be pinned to the

cork in the way figured, leaving the head of the pin sufficiently above the insect to grasp with the fingers.

Care must be taken not to have the insect too far down on the pin, as its legs in that case would touch the bottom of the box, and break off. Insects may be killed by immersing them in alcohol for a few minutes.

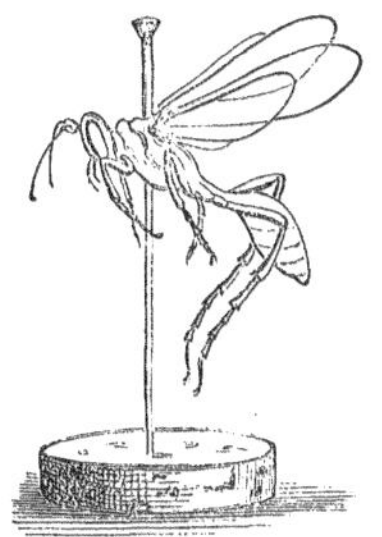

FIG. 51.—INSECT PINNED

Butterflies may be killed by compressing the body between the thumb and forefinger, as shown in the figure, using just force enough to kill, without crushing them. The fumes of benzine, or ether, and of certain poisons, will also kill insects, but these substances should not be suggested to young pupils, as their use is dangerous. (Teachers will here use their judgment according to the character of their classes.) The rude box and common pins are offered simply for experimental collections. The ingenuity of a pupil, where neater collections are desired, will readily secure better ways of making them.

49. The pupils may go out in a class in quest of material

for study, and this is the best way, as the activity and success of one will act as a stimulus to the others.

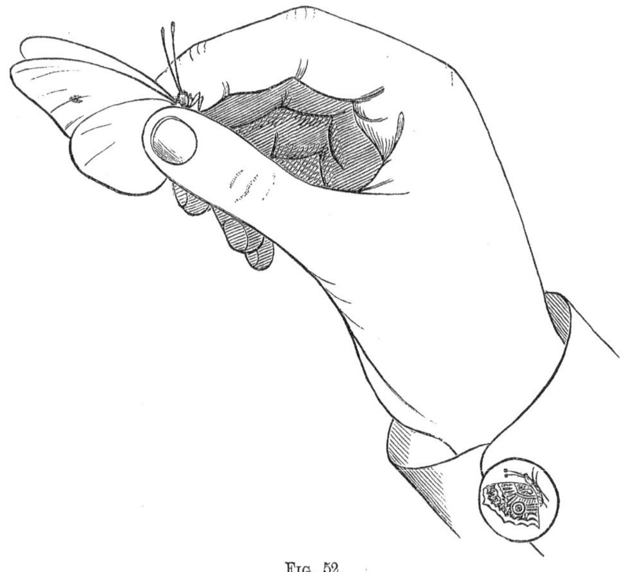

Fig. 52.

In the country, the best places to collect are by the roadsides, or borders of woods and groves; in the gardens, and by the fences, or along the shores of lakes and brooks, under stones and stumps, the bark of fallen trees, or beneath the layers of dead leaves. Insects are scarce in deep woods, and in large, open tracts of pasture-land.

In the cities, the parks and gardens will afford good collecting-grounds, as under plank-walks and boards many insects find shelter. Alongside of railroad-tracks the discarded sleepers often hide many a curious beetle. In short, let the pupil peer under any object large enough to afford shelter to these creatures. By following the furrows made by a plow, certain kinds will surely be met with. The pupil must be urged to pick up every thing that he thinks is an insect, such as grasshoppers, beetles, flies, ants, spiders, etc.

In a single holiday afternoon the pupil will have gathered some of the following animals:

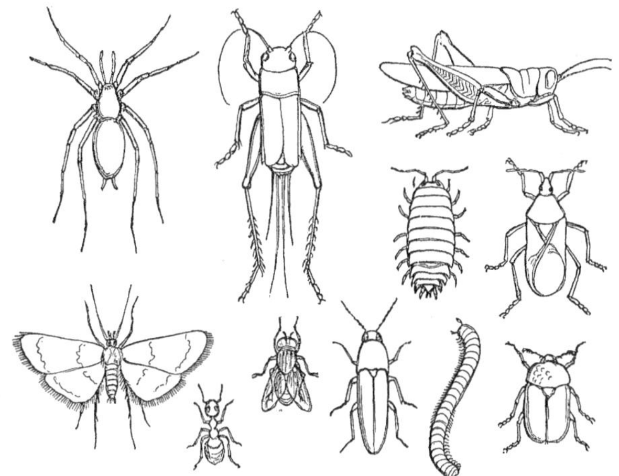

FIG. 53.—SOME OF THE ANIMALS COLLECTED.

CHAPTER VIII.

PARTS OF AN INSECT.

50. THE animals are now to be carefully examined. Let the pupils pick out, and arrange together in one portion of the box, all of those which have three pairs of legs. In

some, the legs will be closely drawn to the body, but by sharp looking they will be found.

After studying these carefully, the pupil will observe that those insects which have three pairs of legs have the body divided into three regions, or parts, called respectively the *head*, *thorax*, and *abdomen*, and that, with few exceptions, they all have wings. Insects having these characters are called *Insects proper*. They are also called *Hexapods*, a word meaning six legs.

These are to be studied first. The other animals collected may be saved for future study.

51. Some insects have the three parts of the body distinctly separated, as in the ants, flies, and wasps. In other insects the parts of the body are very close together, so that it is difficult to distinguish the dividing line, as in certain beetles. Let the pupils examine each insect, and make out the head, thorax, and abdomen.

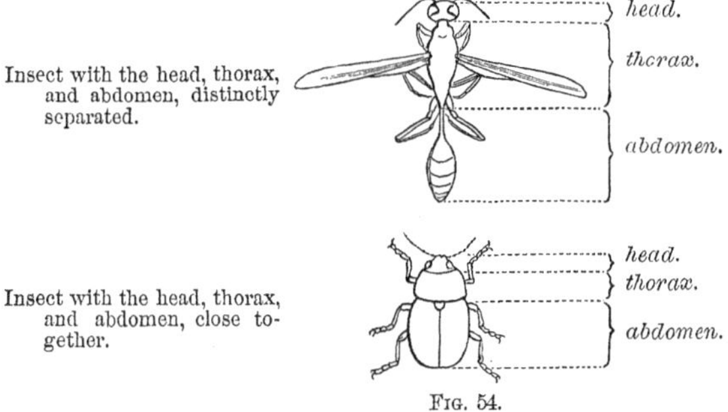

Fig. 54.

In the head, we find the *mouth*, the *eyes*, and the *feelers*, or *antennæ*.

The mouth is on the under side of the head, and is surrounded by certain parts called *mouth-parts.* These parts differ greatly in different insects.

52. In those insects that chew their food, such as the beetles and grasshoppers, certain of the mouth-parts act as teeth, or jaws, and, being joined to the right and left sides of the mouth, move sideways, and not up and down, as in other animals. In other insects some of the mouth-parts are very long and slender, so as to form a long, sharp sting, as in the bugs proper, so that they use them to suck the juices of plants upon which they feed. Or, the parts are again modified in shape to form a long, slender tube, by which the nectar of flowers may be sucked, as in the butterflies.

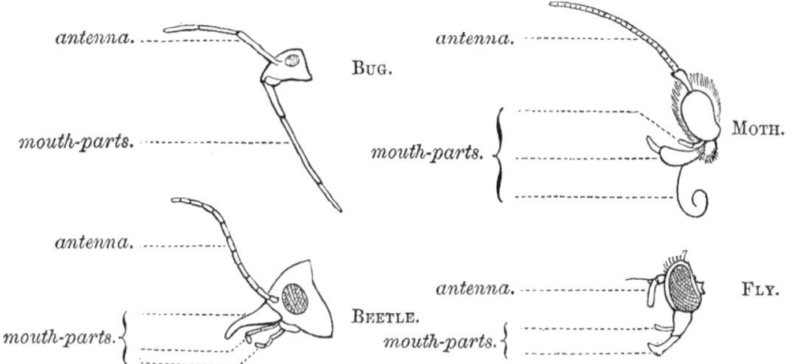

Fig. 55.—Showing Mouth-parts of a few Insects.—The Heads are separated from the Bodies, and are facing the Left, and drawn in Profile as seen from their Left Sides.

In the common house-fly, the mouth-parts appear as a proboscis, a kind of fleshy appendage which is bent up when not in use. When the fly feeds, the proboscis unbends, and the food is lapped up by it. Let the pupils carefully watch a fly as it feeds upon a bit of sugar, or as it laps the hand.

In the butterfly and moth the pupil may uncoil the long tongue with a pin. It resembles in appearance a watch-spring.

53. On the front of the head are two horns, or feelers, called *antennæ.*

These are variously jointed, and vary greatly in different insects.

In butterflies, they are generally long and slender, and swollen at the tips, like a drum-stick. Sometimes they are thread like, and in others the antennæ are barbed on the sides, and look like a feather, as in certain moths. In some beetles they are strongly jointed. In the common house-fly, they hang down in front of the head.

Below are given figures of the left antenna of several different insects showing how different they are in different kinds of insects. The pupils might save the antennæ of different insects and glue them to a card, writing opposite each one the name of the insect, whether fly, beetle, or locust.

54. On the sides of the head are round, smooth places, and these are the eyes. They are entirely different from the eyes of most animals, for, when examined under the microscope, they are seen to be divided into little spaces, looking very much like the surface of honey-comb. Each of these little spaces represents a separate eye. Some insects have hundreds and even thousands of these little spaces, or eyes. For this reason, such kinds of eyes are called *compound eyes.*

Under the microscope three minute black dots may be

seen on top of the head between the compound eyes, and these are called *simple eyes*.

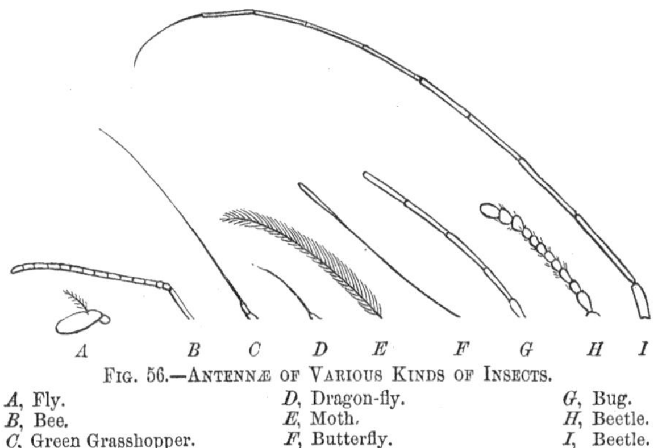

FIG. 56.—ANTENNÆ OF VARIOUS KINDS OF INSECTS.

A, Fly.
B, Bee.
C, Green Grasshopper.
D, Dragon-fly.
E, Moth.
F, Butterfly.
G, Bug.
H, Beetle.
I, Beetle.

In many insects, as in the dragon-flies for example, the compound eyes are very prominent and cover the sides of the head, enabling the insect to look backward as well as forward.

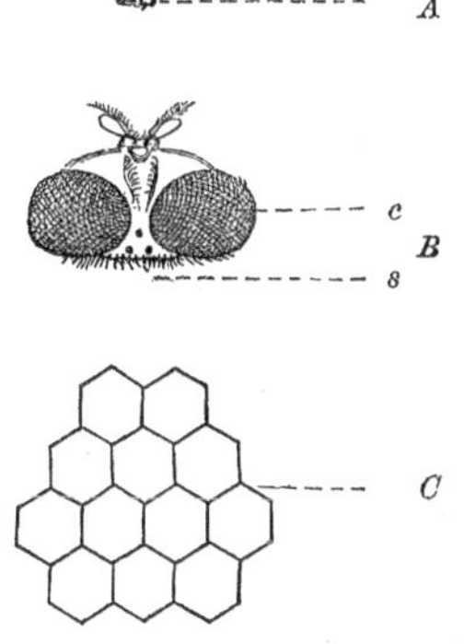

FIG. 57.—SHOWING COMPOUND AND SIMPLE EYES.

A, Head of Fly, natural size.
B, Head of Fly enlarged eight times; *c*, compound eye; *s*, simple eyes.
C, Portion of the surface of a compound eye highly magnified.

55. Thus far we have learned that an *insect proper*, or *true insect*, has the body divided into three parts or regions, called the head, the thorax, and the abdomen.

Let the pupil take a dead fly, and first pull off carefully the legs and wings, and afterward separate the head from the thorax, and the thorax from the abdomen. Having separated the parts in this way, they may be pasted to a card in this manner, writing the correct name beside each part, or region, as shown in Fig. 58.

Fig. 58.—Card, with Regions of an Insect glued to it, and marked.

The principal parts of the head are the mouth-parts, compound eyes, simple eyes, and antennæ.

56. In studying the thorax, the pupil may select some common insect (a large fly, or a bee, will answer the purpose), and pull off the head and abdomen. A common house-fly separated in this way may be stuck upon a card. By experimenting with a number of insects in this manner, the pupil will soon learn that insects not only have the body divided into three sections, but that the thorax invariably has attached to it the legs and wings—the legs being at-

tached to the under side of the thorax, while the wings are attached to the upper side thereof.

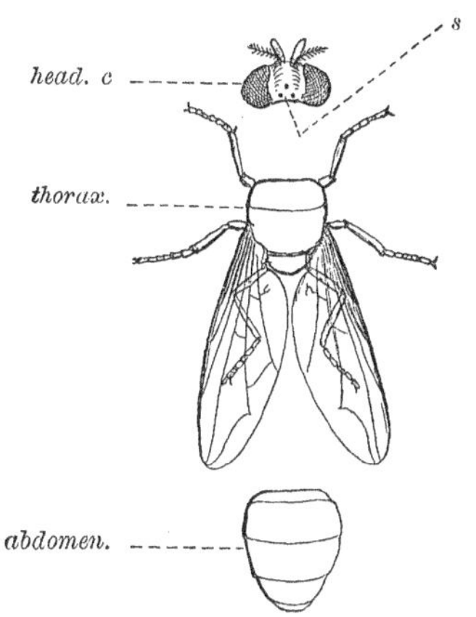

FIG. 59.—Head, having mouth-parts, antennæ, compound eyes, *c*; and simple eyes, *s*.
Thorax, having legs and wings.
Abdomen, never having legs or wings, but having certain appendages at the extremity.

57. The wings of insects are never more than four in number, and these are arranged in two pairs.

The group of insects to which the house-fly belongs has but two wings, or a single pair, and among this group (and other groups of insects as well) there are some which have no wings.

The wings are very different in shape and structure in distinct kinds of insects. In the common fly they are quite small, and transparent. In the butterflies they are large and broad, and are covered with minute scales which rub off on the fingers like dust. In the dragon-fly the wings are long and narrow.

In all these wings the pupil will observe a net-work of lines, which stiffen the wing and support the delicate membrane constituting the wing, just as the frame of a kite stiffens and supports the paper that is stretched upon it. These lines are called *veins*, or *nervures*. To study the *vena-*

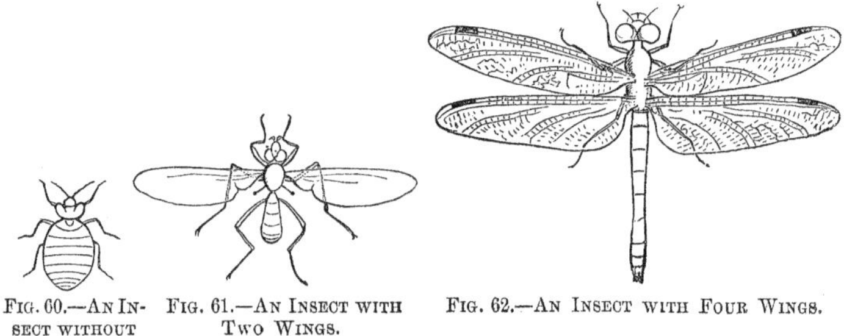

FIG. 60.—AN INSECT WITHOUT WINGS. FIG. 61.—AN INSECT WITH TWO WINGS. FIG. 62.—AN INSECT WITH FOUR WINGS.

tion of the wings, is to study the way in which these veins are arranged. It would be well for the pupils to stick upon a card a number of different kinds of wings, such as those of the grasshoppers, beetles, flies, wasps, and label them accordingly.

58. In many insects the forward and hinder pair of wings are of the same nature, as in the butterflies, moths, bees, wasps, and dragon-flies. In other insects, however, the forward-wings differ in character from the hind-wings. Thus, in the grasshopper the forward pair of wings are more dense in structure than the hind-wings, though the little veins may be seen closely crowded together. They differ as well in form. (*See* Fig. 64.)

In other insects, as in the squash-bug, the front-wings

have the half nearer the body dense and stiff, while the remaining portion of the wing is very thin, or membranous.

In the beetles the front wings are hard throughout, and in most of them are bent and moulded to the shape of the body, and, when closed, form a tight cover over the hind-wings.

The forward-wings of a beetle are so unlike ordinary wings, that they are not called wings, but are known as *elytra*, a single one being called an *elytron.*

59. When insects are at rest, they generally bring their wings into a position different from that taken by them in flight. In certain dragon-flies, however, the wings when

Butterfly at rest with wings meeting over the back.

Moth at rest with the wings sloping on the sides of the body.

FIG. 63.—INSECTS AT REST.

at rest assume the same position as they do when flying. In the butterfly the wings are brought together over the back when at rest, while the moths rest them sloping over the abdomen, the front-wings covering the hinder-wings.

In the grasshopper, the front-wings are long and narrow, while the hind-wings are large and broad. When the grasshopper is at rest, the hind-wings are folded together precisely like a fan, and, when closed, rest against the sides of the abdomen, the long, narrow front-wings closing down upon them, and covering them.

60. Let the pupils prepare a grasshopper, with the wings spread as in the act of flying. A specimen which is dry may be moistened by wrapping it up in a piece of wet cloth, and letting it remain a day or two.

FIG. 64.—GRASSHOPPER WITH THE WINGS OF ONE SIDE EXPANDED.—*f*, Forward-wing; *h*, Hinder-wing.

Having softened the joints of the insect in this way, it may then be pinned to a piece of cork, or a pin-cushion, and, the wings having been stretched, they may be pinned in this position, using triangular bits of card through which the pins are passed to hold the wings in place, as represented in Fig. 64, which shows a grasshopper with the wings on one side of the body pinned in the way described. When the insect becomes perfectly dry the wings will remain in the position in which they were pinned.

A common beetle should be prepared in the same way.

In the beetle the front-wings are very hard and are closed tightly over the hind-wings. With a pin, or the blade of a knife, the upper or front wings may be opened, and beneath these will be seen the hind-wings, not folded like a fan as in the grasshopper, but folded or bent in the middle, as the arm is bent at the elbow.

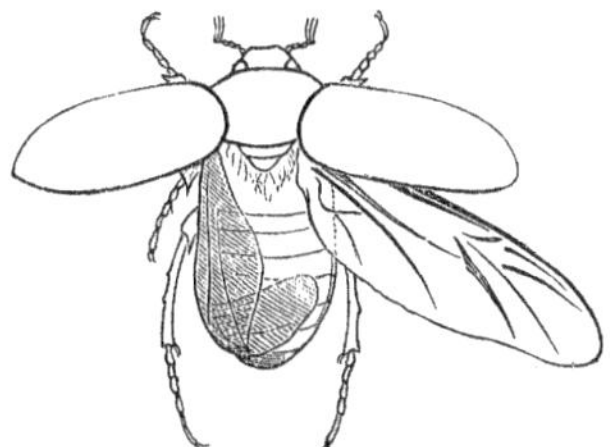

Fig. 65.—A Beetle with the Elytron and Hind-wing of the Right Side open, and the Elytron of the Left Side open with the Left Hind-wing folded in its Natural Position when closed.

61. The abdomen has no wings or legs, but is plainly marked with lines running across the abdomen transversely.

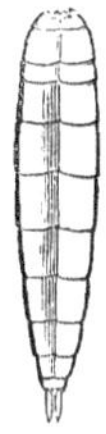

Fig. 66.—Abdomen of a Dragon-Fly, showing Rings or Segments.

These lines show the separation of the abdomen into rings, or segments. In insects with lengthened and slender abdomens the segments are long, and the abdomen, when bent or curved, bends at these joints, as shown in Fig. 67.

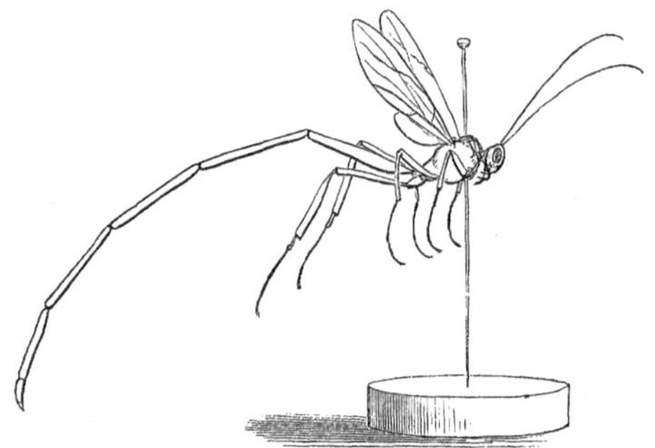

FIG. 67.—INSECT WITH A LONG, SLENDER ABDOMEN.

If the pupil can handle these parts delicately enough, he may be able to separate the abdomen at these joints, into a series of rings, or segments, and glue them on a card, marked "*Rings or segments of the abdomen.*" In the grasshoppers the segments show very plainly. On the hinder part of the abdomen there are various appendages, sometimes so short as to be scarcely perceptible, sometimes long, and thread-like, as in the May-fly (Fig. 98); again, in the shape of a sharp sting, as in the hornet. In the cricket, they are quite long and conspicuous. These appendages vary greatly in different insects.

CHAPTER IX.

PARTS OF AN INSECT (CONTINUED).

62. THE pupils have learned that the abdomen is divided into rings or segments, and the division between these segments is plainly seen in most insects.

The thorax is divided in a similar manner, only the lines which divide the thorax are not so plainly seen.

The number of segments in the thorax is three. To the first segment, the head and first pair of legs are attached; to the second segment, the second pair of legs and the first pair of wings are attached; and to the third segment, the hind pair of legs, the hind pair of wings, and the abdomen, are attached. The three segments of the thorax have special names: the *prothorax*, this being the forward segment, next to the head; *mesothorax*, being the middle segment; and *metathorax*, being the last segment. Arranging these segments with the appendages attached to them in a table, they would appear as follows:

THE THORAX IS COMPOSED OF THREE SEGMENTS.	1st Segment, *Prothorax*,	Has attached to it the first pair of legs.
	2d Segment, *Mesothorax*,	Has attached to it the second pair of legs and the first pair of wings.
	3d Segment, *Metathorax*,	Has attached to it the third pair of legs and the second pair of wings.

63. Let the pupils now endeavor to dissect a beetle, carefully separating the segments of the thorax, and, if possible, the minute jaws and other mouth-parts, and stick them on

a large card, with the names of the different parts neatly marked upon the card, as in the accompanying figure (Fig. 68):

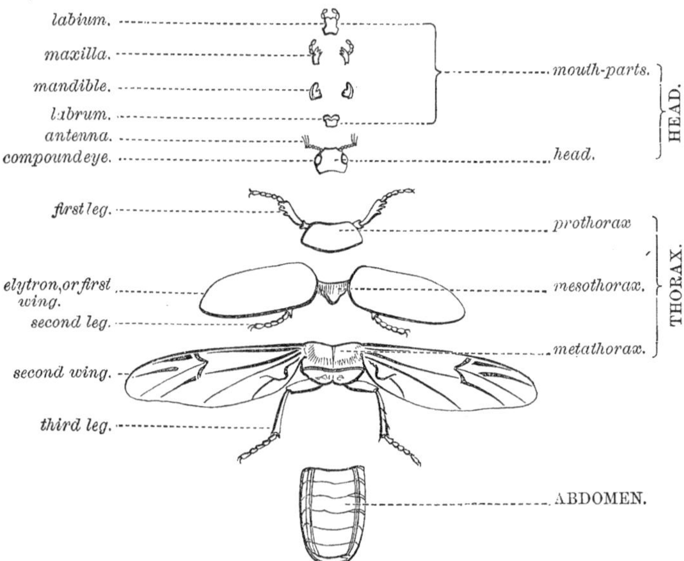

FIG. 68.—A COMMON BROWN BEETLE, WITH THE PARTS SEPARATED.

Having now learned something about the parts of an insect, and having seen how much these parts vary in size and appearance in different insects, the pupils are better prepared to understand the surprising modification which the *mouth-parts* undergo in the different groups.

64. The parts of an insect's mouth, generally speaking, consist of an upper lip, called the *labrum;* a pair of jaws, called *mandibles;* a pair of smaller jaws, called *maxillæ*, to which are attached little jointed feelers, called *maxillary*

palpi; and a lower lip, called the *labium*, which represents still another pair of jaws joined together; to this joined piece, or *labium*, are attached a pair of jointed feelers, called *labial palpi.* In Fig. 68 these parts are shown separated from the head.

The numberless varieties of mouth-parts, peculiar to different insects, are in reality made up by modifications of the parts above described. Thus, in one group of insects, the mandibles may be lengthened out into a piercing-like sting, while some of the other parts may be reduced in size, or become almost obsolete. In another group the maxillæ may be greatly elongated, with their edges joined to form a tube, while the other parts of the mouth are scarcely to be discerned. In another group the labium may be greatly lengthened to form a tongue-like organ for lapping up food, while the mandibles—so big and hard in some insects—are barely perceptible, and of no use to the insect.

Not only, then, do these parts assume different proportions and different shapes in the different groups of insects, but they also vary greatly in being very hard or very soft.

If the pupils will examine the different kinds of insects' wings, taking the front-pair of wings for example, they will find a marked difference between them, some being very large and transparant, as in the dragon-fly, others being hard and opaque, as in the front wings or elytra of a beetle. Compare the broad and brilliant-colored wing of the butterfly with the straight and narrow fore-wing of a common grasshopper. And yet these are all wings. In a similar way do the mouth-parts of an insect vary.

65. In the head of a mosquito, what appears to be a single sting, by which the animal pierces the flesh and sucks the blood, is in reality composed of long, delicate, thread-like parts, which represent the mandibles, maxillæ, and the tongue, or *ligula*, which represents a prolongation of the labium. In the bugs the mouth-parts are compacted into a hard beak—the *piercer*, so called, consisting of mandibles, maxillæ, and labium, the labrum being represented by an acutely triangular piece.

The mouth-parts of a beetle have already been described in general terms. They are represented as separated from the head in Fig. 68, while in Fig. 69 a side-view of another beetle is given in which the mandible shows very prominently, while the labrum, labium, and maxillæ, do not show at all, as they are concealed by the other parts. The maxillary and labial palpi are seen, however.

66. In the butterfly the labial palpus is seen very large and prominent, while the coiled, thread-like tongue represents the pair of maxillæ lengthened and joined, forming a long elastic tube which can be coiled or uncoiled by the insect, and through which it sips the nectar of flowers. In the mouth-parts of a house-fly the parts are soft and fleshy, and united together to form a sort of proboscis; the maxillæ are minute; the maxillary palpi are present as simple jointed appendages; the mandibles are minute, and useless; while the labium is greatly developed, having a broad end which is

divided into two lobes at the extremity, by means of which the fly laps up its liquid food. The insides of these lobes are rough, and the irritation which flies produce when they alight upon the hand is caused by the scratching of these rough surfaces.

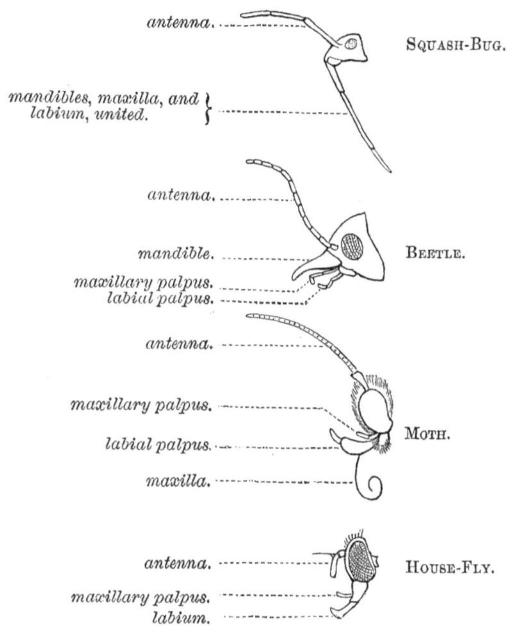

FIG. 69.—HEADS OF A FEW INSECTS SEEN FROM THE SIDE, SHOWING MOUTH-PARTS, NAMED.

67. Much more may be learned about the mouth-parts of insects, and the pupils might attempt the separation of the mouth-parts of such insects as the grasshopper, beetle, wasp, and butterfly, sticking these parts, when separated, upon a piece of white card, as shown in Fig. 68.

It will be advisable, also, for the pupils to utilize the broken specimens of insects by selecting the wings of different insects and gluing them upon a card, labeling each one. Different kinds of antennæ might be fastened upon another card, and the legs of some widely-different insects may be arranged in the same way. For example, let them take the hind-pair of legs of a grasshopper and of a water-beetle. It will be instructive to observe how different these two kinds are, and how admirably one is adapted for jumping, while the other is so perfectly fitted for a paddle. By making comparative collections of this kind the pupils will learn a great deal regarding the structure of insects.

68. The pupils have thus far learned that a true insect has the body divided into regions called the head, thorax, and abdomen; that the head bears the mouth-parts, antennæ, and eyes. The thorax has the legs and wings, while the abdomen has only the caudal or tail appendages, and these are not often apparent. They have also learned that the thorax is made up of three segments, not often plainly marked, while the abdomen is composed of a greater number of segments, in most cases very plainly apparent. As each segment of the thorax is characterized by having attached to it a pair of appendages, and as the head contains a number of appendages, it is believed by many naturalists that an insect's head is composed of a number of segments, so closely merged together, however, as not to be distinguished, except theoretically. As naturalists, however, differ in their estimate of the number, we will leave this difficult problem for more advanced students to study.

CHAPTER X.

GROWTH OF INSECTS.

69. As the study of the growth of an insect, from the egg to maturity, requires some time and considerable care, the different stages of such growth may be described and taught with what examples the pupil may be able to collect.

An afternoon may be spent exclusively in collecting the following objects:

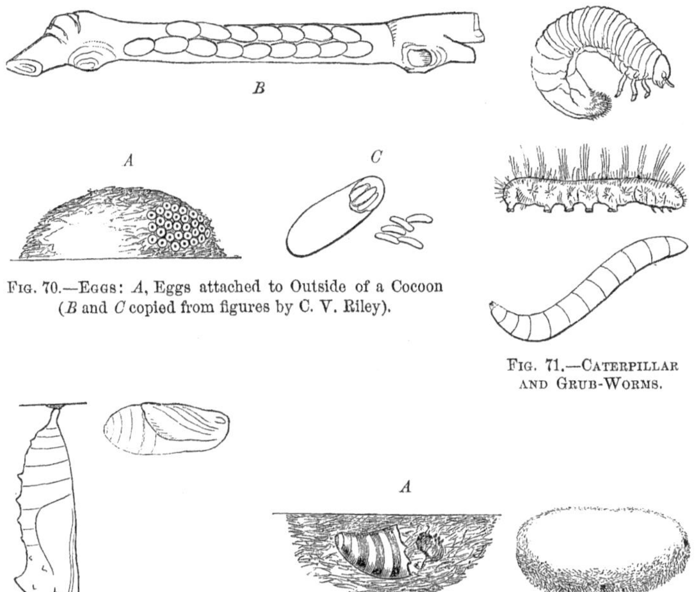

FIG. 70.—EGGS: *A*, Eggs attached to Outside of a Cocoon (*B* and *C* copied from figures by C. V. Riley).

FIG. 71.—CATERPILLAR AND GRUB-WORMS.

FIG. 72.—CHRYSALIDES.

FIG. 73.—COCOONS: *A*, showing Inside of Cocoon, containing the Remains of a Chrysalis-Skin.

In the spring and fall the eggs of the canker-worm moth may be found in abundance on fences and trees in cities. They are very minute, and are found in clusters arranged like stones in a pavement, but with greater regularity. With a sharp knife a shaving of wood may be cut off, taking the eggs with it. (*See* eggs in Fig. 77, *b*.)

If they are collected in the spring-time, little worms will hatch from them in the course of a few weeks, and these may be fed on the young leaves of the elm-tree.

Eggs of other insects may be found on fences, leaves, and twigs of plants; also on the leaves of the squash-vine, and other plants in the garden.

Certain eggs may be found upon the twigs of apple-trees, covered with a shiny coating, like varnish. For all these objects the pupils will have to hunt carefully, as only the keenest eyes will find them out.

70. *Caterpillars* and *grub-worms* are found everywhere, so common indeed that the pupil has only to examine the fences as he goes to school to secure some. For certain kinds of grub-worms, he may dig in the garden, follow the furrow made by a plough, or tear the bark from some dead tree, and discover the specimens he is in search of. For *chrysalides* and *cocoons* the pupils may be directed to search on old garden-fences, under stones and dry boards.

Some chrysalides will be found hanging with the large end downward, as in Fig. 72; others will be found suspended by the small end, with a delicate thread around the middle, holding the chrysalis horizontally, or vertically, against the fence, as in Fig. 80, *C.*

Some of them are encased in a mesh of threads, which may be built against the fence, or under the edges of clapboards on the sides of houses. And within the cocoons collected, the chrysalides, or their empty cases, will be found, as in Fig. 73.

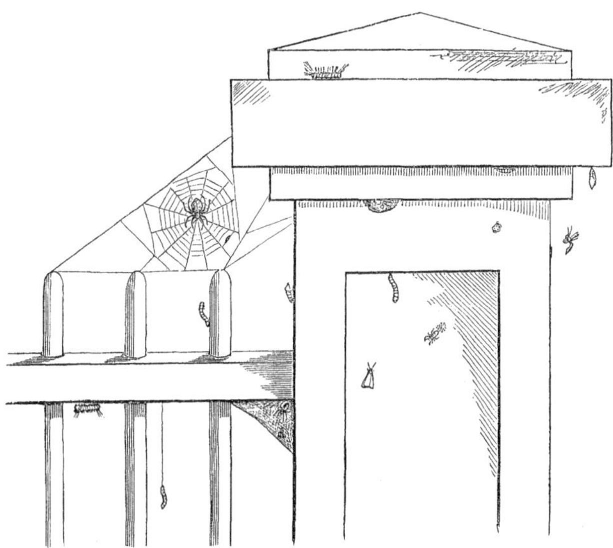

FIG. 74.—PORTION OF A FENCE, HAVING UPON IT, AMONG OTHER THINGS, EGGS, CATERPILLARS AND CHRYSALIDES OF INSECTS.

Rich collecting-places may always be found on old garden-fences in cities and towns. Fences surrounded by trees and bushes will oftentimes have a great many insects lurking under projecting edges—caterpillars climbing up the posts, and chrysalides attached to the rails of the fence. Fig. 74 shows a portion of a fence of this kind.

With the eggs, caterpillars, and chrysalides on hand, the pupils are ready to study the life-history of an insect.

71. Many animals, as fishes, snakes, and birds, lay eggs, and from these eggs little creatures are hatched that resemble the animals which laid the eggs.

Insects also lay eggs, but from these eggs come little creatures which do not at all resemble the insect that produced them.

The different kinds of eggs collected by the pupils were laid by insects of different kinds. For example, if they have collected eggs like the following (Fig. 75), the insect that laid them was a moth, and looked like Fig. 76.

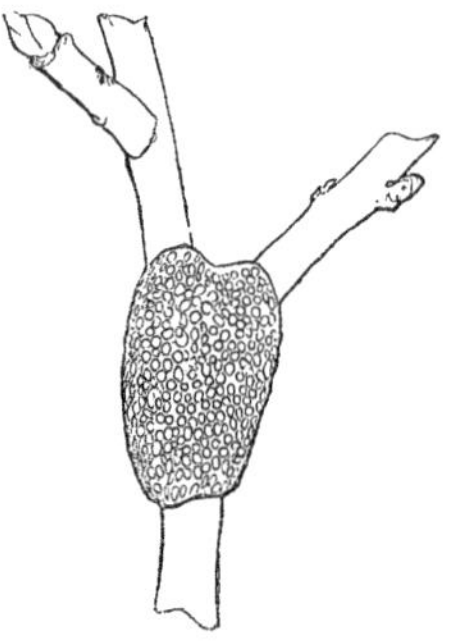

FIG. 75.—EGGS LAID ON THE TWIG OF AN APPLE-TREE.

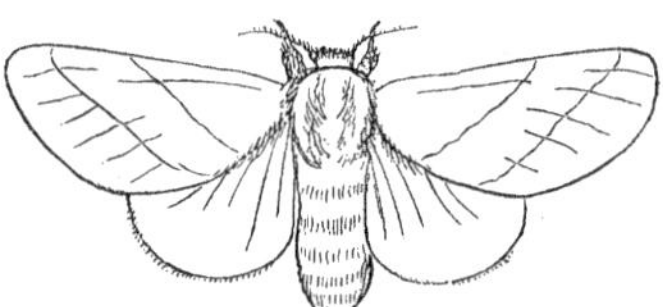

FIG. 76.—INSECT WHICH LAID THE EGGS IN FIG. 75.

Now, if the pupil will keep these eggs in a box, there will hatch from them little animals resembling worms, very tiny at first, but growing rapidly if supplied with appropriate food. Insects generally lay their eggs in such places that the worm, or caterpillar, coming from them can easily find access to its proper food, and this food in the case of most caterpillars consists of leaves, or the wood, bark, or juices, of plants and trees.

72. Commencing with the egg, the pupil should get, if possible, the insect while in the act of depositing her eggs, and this will not be difficult to do in the case of the canker-worm moth, whose caterpillars commit such ravages upon the elm-trees.

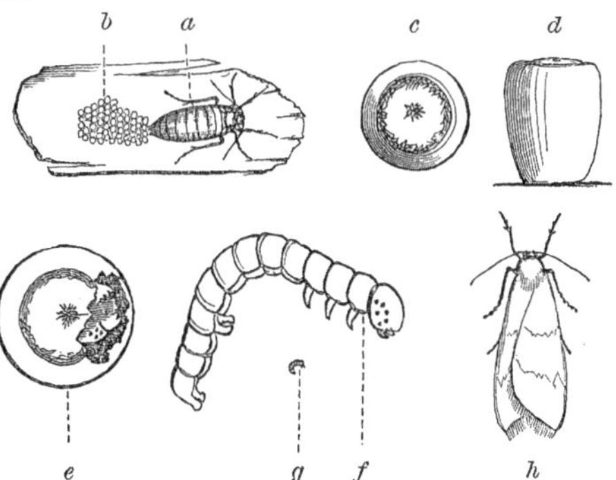

FIG. 77.—CANKER-WORM MOTH, EGGS, AND WORMS: *a*, Female Canker-worm Moth laying her Eggs, *b*; *c*, Top-View, and *d*, Side-View of an Egg magnified; *e*, Canker-worm eating its way out of the Egg, magnified; *f*, Magnified View of Canker-worm; *g*, Natural Size of Canker-worm after leaving the Egg; *h*, Male Canker-worm Moth.

The female of the canker-worm moth has no wings. They are very common in early spring and in the fall, laying their eggs on fences. Fig. 77, *a*, represents the female moth depositing the eggs; *b*, the eggs being deposited upon a chip which was cut from a fence while the female was at work; *c*, represents a top view of one egg magnified; *d*, represents a side-view of the same egg; *e*, represents another egg with the canker-worm eating its way out; *f*, represents the canker-worm highly magnified, after it has crawled out from the egg; *g*, shows the natural size of the worm; *h*, shows the appearance of the male canker-worm moth. The female moth which is laying the eggs differs from the male in having no wings.

Now, if fresh elm-leaves are placed in the box with the worms, they will commence to feed on them. The eggs hatch out just as the leaves commence to grow, and consequently the young worms have tender leaves to feed on at the outset.

The worm grows rapidly, and after a few weeks ceases feeding, and, dropping to the ground, or lowering itself down by a thread spun from the head, buries itself just below the surface of the ground, and there changes into a chrysalis, forming a rude cocoon of earth about it. At the proper time there comes from the chrysalis a male canker-worm moth with wings, or a female canker-worm moth without wings.

73. From the eggs, then, come worms or caterpillars. The worms or caterpillars change into chrysalides, and sometimes these are inclosed in cases or cocoons. From the chrysalides comes the perfect insect, similar to the insects which first laid the eggs.

If it is desired to keep the caterpillars alive, the pupil should try to find them when they are feeding, and to observe the kind of leaf they are eating, and then, by giving them fresh leaves of the same kind as they need them, the caterpillars will grow rapidly, and ultimately change into chrysalides. It is a common thing to see boys collect caterpillars and place them in a box, with grass to feed upon; the grass dries, and of course becomes unfit for food, and even if it were fresh the caterpillar would not eat it, unless it were its natural food. If the pupils wish to raise caterpillars, they must be sure and furnish them with the kind of leaf they are accustomed to feed upon. It may be an elm-leaf, or a cabbage-leaf. Thus, there is a common worm which they will find feeding on the leaves of the Tartarian honeysuckle, as in Fig. 78. To raise this worm, therefore, it is necessary to furnish it, from time to time, with the leaves of this honeysuckle.

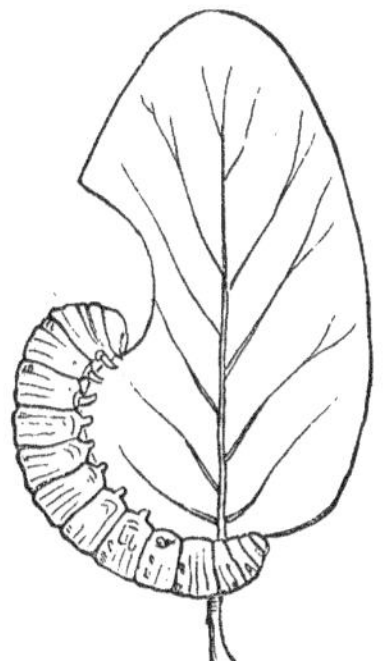

Fig. 78.—Worm feeding on the Leaf of the Tartarian Honeysuckle. (Copied from figure by J. H. Emerton, in Packard's "Guide to the Study of Insects.")

In its growth, the caterpillar usually sheds its skin three or four times.

After the caterpillar has become full-grown, it stops eating, and crawls about in a restless manner.

In the summer and fall, various kinds of caterpillars are seen crawling over the sidewalks and along fences. They are not now searching for food, but are seeking an appropriate place where they can change into the chrysalis state.

From the long, active, and often devastating caterpillar, having three pairs of small legs in front, and other pairs of blunter legs behind, the creature gradually changes into a body, blunt in front, tapering behind, with no indications of legs, head, or any of the appearances seen in the caterpillar, except that the hinder part still shows the division of that portion into rings or segments as in the caterpillar, and signs of life are still manifested by this portion moving from side to side, when touched. Many caterpillars spin a case or cocoon, as it is called, in which it incloses itself previous to changing to a chrysalis. The thread with which they make this case issues from a little tube in the lower part of the mouth or labium. Silk is made from the thread composing the silk-worm's cocoon.

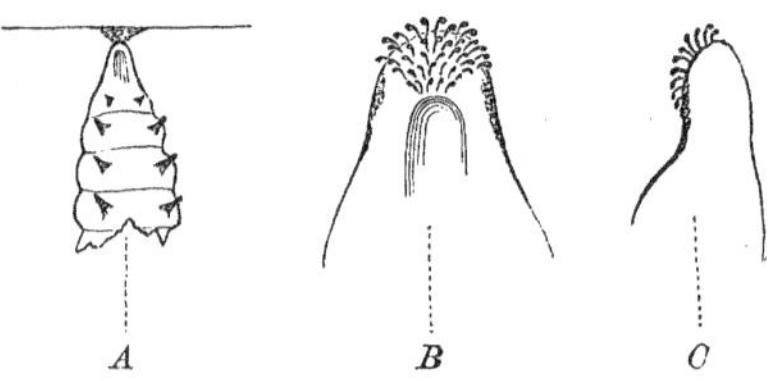

FIG. 79.—*A*, Hinder Portion of Chrysalis-skin hanging; *B*, *C*, Enlarged View of Hinder End, to show the little Hooks by which it hangs.

74. Fig. 79, *A*, shows the hinder portion of a chrysalis-skin, the insect having escaped from it, and the forward part

having fallen to the ground. The chrysalis is seen suspended by the tail, and is held there by little hooks on the end of the tail, which become entangled in a sort of web previously made by the caterpillar; *B* and *C* represent different views of the chrysalis-tail enlarged so as to show the hooks. Pupils will be sure to find these empty chrysalis-skins attached under the projecting portions of fences.

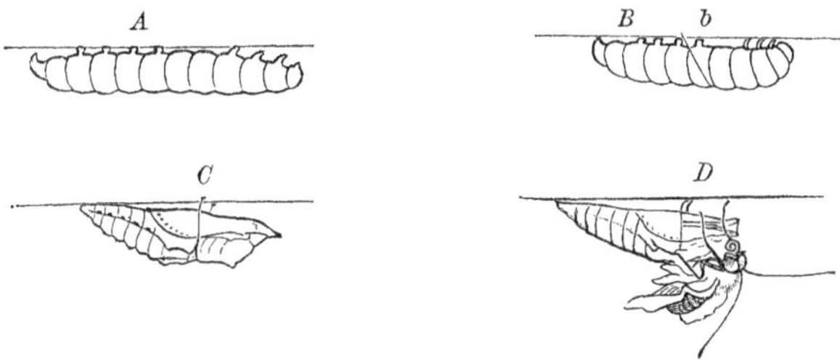

FIG. 80.—*A*, Caterpillar getting ready to change into a Chrysalis; *B*, Just ready to shed its Skin previous to changing; *b*, little Band to hold it up; *C*, Chrysalis; *D*, Butterfly just escaping from Chrysalis, the Wings just being unfolded.

Fig. 80 shows different stages of a cabbage-worm, from the worm stage to the chrysalis stage: *A*, representing the worm as it assumes a position under the projecting edge of a fence; *B*, after it has supported itself round the body by a delicate thread, *b*, and attached itself by the tail at the same time; and, *C*, representing its complete chrysalis condition; *D*, shows the butterfly just escaping from the chrysalis, the wings still being rumpled. After having escaped from the chrysalis, the butterfly generally clings to the empty case till the wings have expanded and dried, when it flies away. Fig. 80 represents a cabbage-butterfly introduced from Europe, and now common in certain parts of New England. Its wings are yellow, with two blackish spots on the forward wing, and one on the hinder wing. The chrysalides of this species are very common on fences, and, when collected in the fall, may be kept through the winter. During February and March the butterflies will come out, and these may be fed on honey or sugar mixed with water, and in this way may be kept alive for some time.

75. The caterpillar, then, having changed into the chrysalis, remains in this condition a few weeks, or even many months, and then the skin slowly cracks open, and out crawls a creature no longer like a caterpillar, but having three pairs of long, jointed legs, the body divided into three very distinct regions—the head, thorax, and abdomen—the thorax having wings, and the head furnished with long antennæ, and provided with mouth-parts suitable for sipping nectar, and no longer like the heavy jaws of the caterpillar, suited only to chewing coarse leaves; in short, a creature resembling the insect which first laid the eggs from which the caterpillar came.

76. Other names are given to these three stages of an insect. The worm, or caterpillar, is called the *larva;* the chrysalis is called the *pupa;* while the perfect insect is called the *imago.* These terms are necessary, for without them the proper condition of an insect could not be easily described.

Take, for example, the caterpillar stage of a butterfly: the same stage in a common fly is known by the name of maggot, and in other insects the same conditions are known by the name of borer, grub-worm, and many other terms. If the pupil learns that all these various names describe a similar stage in the lives of these insects, it is much more convenient to have some general term describing all these stages, such as *larva*, or *larval stage.*

77. While most insects pass through changes similar to those above described, there are others, such as the grasshoppers, crickets, roaches, and bugs proper (a group of insects which includes the squash-bug, chinch-bug, and bed-bug, all of which have a disagreeable taste and odor, and to which naturalists restrict the name of bug), which do not pass through a caterpillar and chrysalis state. The young hatch from the egg, and closely resemble the adult insect,

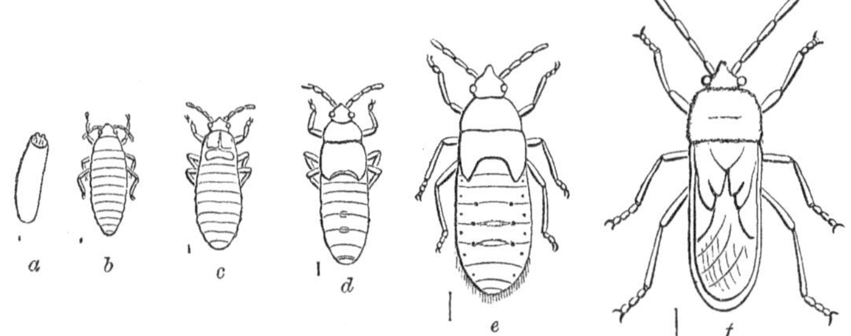

FIG. 81.—DIFFERENT STAGES OF THE CHINCH-BUG: *a*, Egg; *b*, Newly-hatched Larva; *c*, Larva after First Moult; *d*, Larva after Second Moult; *e*, Pupa; *f*, Perfect Insect.

[These figures are copied from the Seventh Annual Report of C. V. RILEY, State Entomologist of Missouri.]

except that it has no wings, and is of course much smaller than the parent. In its growth it moults or sheds its skin, and each moult reveals its wings more advanced in growth, till finally, on the last moult, it attains the size and features of the mature insect. And even in this growth, so unlike the moth and butterfly, the terms larva and pupa are applied to certain stages of their history.

The foregoing figure represents the egg and successive stages of the chinch-bug, an insect which has been so destructive to various crops in the West. The figures are all enlarged; the little line at the lower left-hand side of each figure represents the natural size.

78. Many insects, as the beetles, flies, moths, butterflies, bees, and wasps, pass through complete and distinct changes from their early condition to maturity, as above described. Other insects, as the grasshoppers, crickets, roaches, and bugs, hatch out from the egg, as little six-footed insects, and not as worms, and in their growth do not pass through an *inactive pupa* or chrysalis stage, but slowly acquire wings, and ultimately attain full growth as above stated. Hence these changes are not so completely defined as the changes in the insects first mentioned. For this reason the term *complete metamorphosis* is used to define the mode of growth of the beetles, flies, and other insects having a similar mode of growth; while the term *incomplete metamorphosis* defines the mode of growth of the grasshoppers, crickets, and others.

79. Many of the larvæ of insects look like worms—so much so, indeed, that they are commonly called worms, such as cut-worms, canker-worms, currant-worms, and the like. The pupils have learned that these are not true worms, but only the larval condition of certain insects.

True worms, however, never change into any thing else. Such, for example, is the earthworm, hair-worm, and leech, and worms which live in the sea.

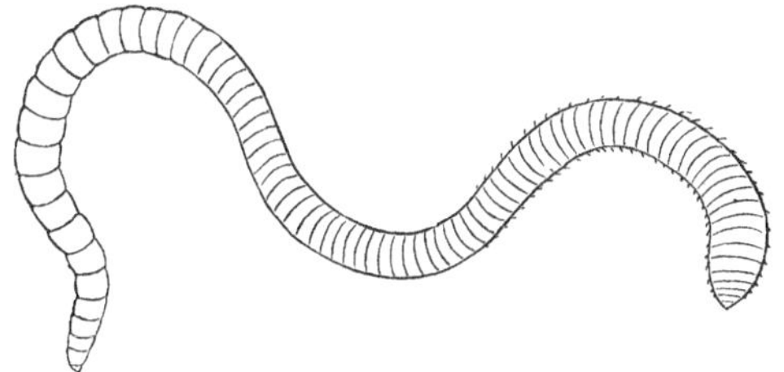

FIG. 82.—EARTHWORM.

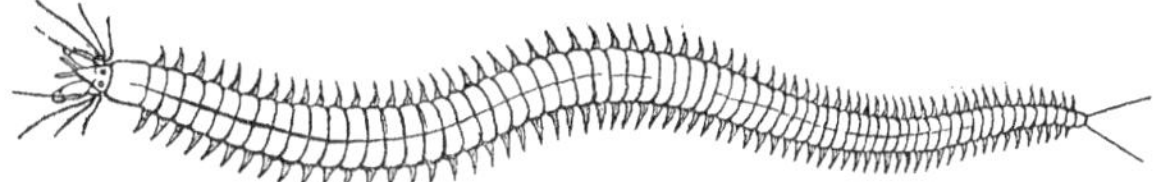

FIG. 83.—MARINE WORM.

These worms, and other true worms, generally speaking, have the body divided into a great many segments or rings, as in the earthworm. In the larvæ of insects, on the contrary, the segments are limited in number. With few exceptions the larvæ of insects have legs, and these legs in the fore-part of the body are jointed. In worms, jointed legs do not occur. The jointed legs of the larvæ number three pairs, and are on the three successive rings back of the head, and consequently correspond to the three pairs of legs in the adult insect.

In certain larvæ other legs occur, but these are not jointed, though often having special structures at their extremities, by which they are enabled to cling. These are called *prop-legs*, or *false legs*.

It would be well for the pupils to collect some leeches and earthworms, and, if they live near the sea-shore, a few worms may be collected under stones at low tide.

Having collected these, let the pupils compare them with the larvæ of insects.

As the larva of an insect comes from the egg, it has its full number of segments at the outset. In the larva of a butterfly, for example, there may be counted, besides the head, twelve segments or rings, and this number does not increase as the creature grows, but remains constant; and, as we have already learned, the creature does not long remain in its worm-like stage, but assumes other conditions, ultimately becoming a creature unlike, in form and habits, the larval condition in which it spent a portion of its life.

80. The true worm, on the contrary, comes from the egg with a very limited number of segments, and as it grows new segments are formed, till in some worms as many as four or five hundred segments are developed before the animal has attained mature proportions, and in this condition it remains; that is, it is complete, never changing or passing through larval or pupal stages to develop into something quite unlike the worm.

Briefly, then, a larva may be distinguished from a true worm generally by its limited number of segments, and, when supplied with legs, having three pairs of jointed ones on the anterior rings of the body.

With few exceptions worms live in the water, and even those which live on the land are limited to damp earth or moist places. While the larvæ of insects are in many cases aquatic in their habits, and breathe or respire by means of gills, the larger number live on the land, feeding on leaves, wood, and substances of a similar nature, and are air-breathers.

CHAPTER XI.

HABITS AND STRUCTURE OF INSECTS.

81. Let the pupils now endeavor to study the habits of certain insects by direct observation. The following sketches are given as aids to the pupils in making independent observations on special insects.

Most insects make no provision for the larvæ, but leave them to take care of themselves, though usually the egg is deposited where the larva coming from it may find proper food at hand.

Other insects prepare cells or cavities in which they deposit their eggs, just as a bird builds a nest to hold its eggs. Certain insects, in preparing these cells, also lay up a store of food ready for the larva when it shall have hatched from the egg. Such is the case with the common mud-wasp. This insect makes a number of little chambers of mud, generally sticking them to the sides of a wall, or to the ceilings of sheds, barns, and attics. These pellets of mud are seen firmly plastered to the wall, rough and irregular in appearance, and at first sight might be mistaken for the work of some mischievous boy.

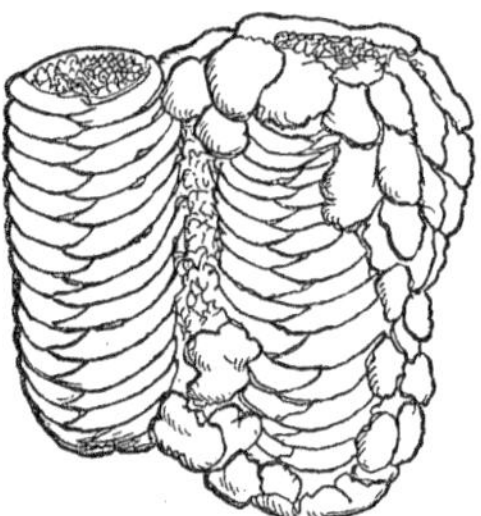

FIG. 84.—A MUD-WASP'S NEST, HAVING TWO CELLS.

Fig. 84 represents their general appearance, for, rough as they appear, on close examination they are seen to be constructed very systematically — the soft clay of which they are built being laid on in alternate layers, looking something like a braid; while the upper portion, being built of looser and coarser material, is put on in irregular lumps. With a thin-bladed knife these mud-cells may be scraped off, and sometimes can be pulled away with the fingers. Upon opening them they will be found either filled with little spiders, or containing yellow-colored larvæ, pupæ in brown skins, or wasps. Their history is as follows:

82. When the mother-wasp gets ready to lay her eggs, she first builds these curious nests of mud, which she collects from the streets, or by the sides of brooks in clayey soil. Having finished one cell, she deposits therein an egg, and then collects a number of small spiders with which she completely fills the cell. After this is done she closes up the top of the

cell with looser mud. Thus she proceeds, constructing cell after cell, going through the same manœuvres with each one. It has been observed that the wasp stings the spiders so as to paralyze, but not to kill them. Hence they remain alive but cannot struggle, and when the egg hatches the little larva coming from it finds in these spiders a store of food on which to feed. These are gradually eaten, and thus room is made for the rapidly-growing larva which, having eaten all the spiders, passes into its pupa state surrounded by its brown chrysalis case, and finally emerges a perfect wasp, when it softens the mud-walls of its nest, by a fluid poured from its mouth, and gnaws its way out.

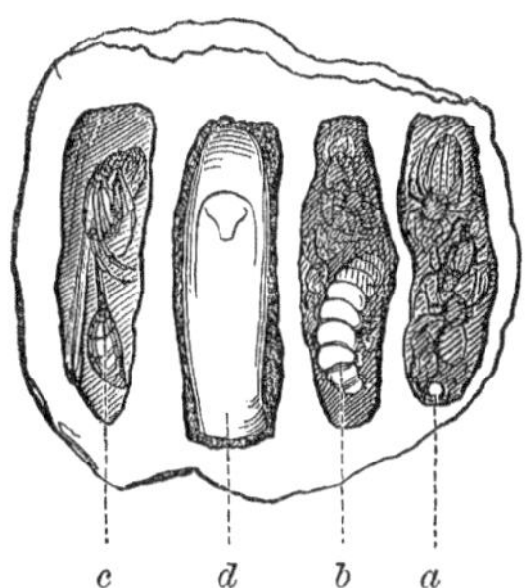

FIG. 85.—SHOWING A NEST OF FOUR CELLS CUT OPEN: *a*, representing a Cell with the Egg at the Bottom, and the remaining Space filled with Spiders; *b*, the Larva full-grown, after having consumed all the Spiders; *c*, the Pupa; and *d*, the Imago, or Perfect Wasp, ready to come out.

Fig. 51 shows one of these mud-wasps pinned.

The pupils may collect these nests or cells in April or May, and by June the wasps will be ready to come out. If collected soon after they are made, the eggs may be found; if a little later, the larva will be found feeding on the spiders; and, still later, the full-grown larvæ and pupæ appear.

In collecting for the cabinet, one nest should be cut open to show the cells and their contents.

83. The mosquito deposits her eggs on the surface of the water, sticking them together in such a way as to form a raft. From these eggs little black creatures hatch, which swim about with a quick, jerking motion. In this condition they represent the larvæ. If the pupils will examine pools and ditches, or even the tubs and barrels of water which often stand about farm-houses, they will be very sure to find some of these animals. They are small and black, and by their

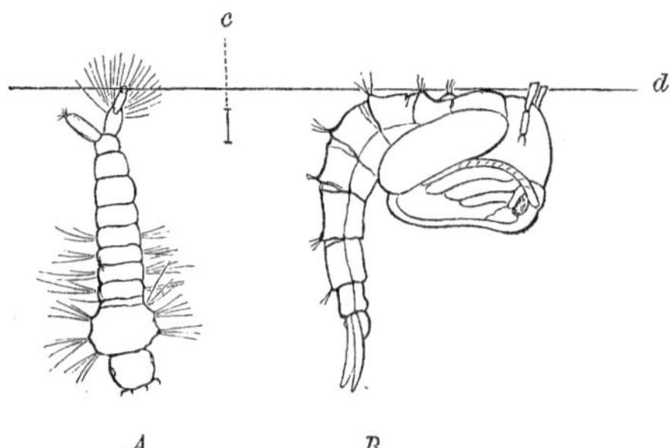

FIG. 86.—*A*, Larva; and *B*, Pupa of a Mosquito; *c*, Line showing Natural Size of Larva; *d*, representing Water-Line.—The larva is seen with the hinder end of the body just projecting above the surface of the water, so that the air may enter through the little tube. The pupa is seen with the back just level with the surface of the water, and through the two tubes, on the back of the thorax, the air is being admitted to the body.

rapid, jerking motion cannot be mistaken. A number may be collected and placed in a jar of water, where all their changes, from the larval to the perfect state, may be watched. They will be seen coming to the surface of the water for air, which they breathe in through openings in the hinder part of the body. Changes soon take place by which they assume the pupa state, and at this time they no longer breathe through the hinder portion of the body, but through two tubes on the back of the thorax. Finally, the pupal skin cracks open, and out crawls the perfect mosquito, for a while resting on the empty pupal case which floats in the water like a raft, and the insect remains supported in this way till the wings become fully expanded and dry, when it flies away.

CHAPTER XII.

HABITS AND STRUCTURE OF INSECTS (CONTINUED).

84. In certain groups of insects the young hatches from the egg, not as a caterpillar, but as a little insect having the body divided into three regions, possessing three pairs of jointed legs, and looking very much like the mature insect, except that it is very much smaller and has no wings.

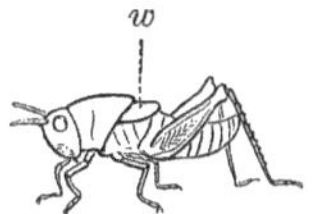

FIG. 87.—YOUNG GRASSHOPPER.—*w*, Wing just appearing.

In the grasshopper, for example, the animal does not pass through a series of abrupt changes, but the creature comes from the egg with the general proportion of the adult insect, save that it has no wings, these growing gradually as the creature increases in age. Fig. 87 represents a young grasshopper with the wings just appearing. At intervals, the insect sheds its skin, or *moults*, the wings continually increasing in length until mature size is reached.

Let the pupils endeavor to collect some young grasshoppers representing various stages of growth, and place these in their collections beside the full-grown one.

By searching in the grass, the cast-off skins of grasshoppers may be occasionally found still clinging to the spears of grass, where they were left when the grasshoppers shed them.

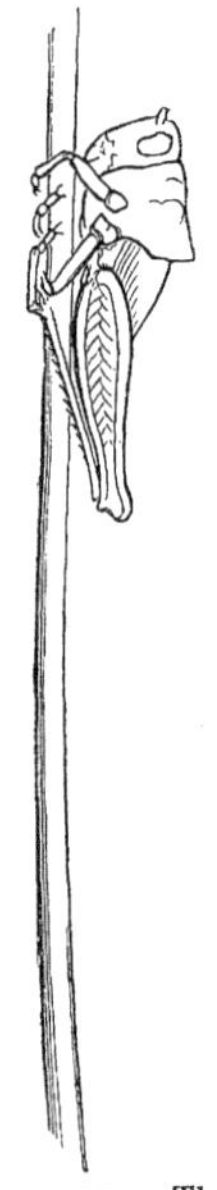

FIG. 88.—CAST-OFF SKIN OF A GRASSHOPPER.—The grasshopper has shed its skin while clinging to a blade of grass. The skin is imperfect, the antennæ and parts of its legs are broken; the abdomen is shriveled, and does not show.

Fig. 88 shows the appearance of one of these cast-off skins.

85. Grasshoppers are often infested with parasites. Frequently the grasshoppers, in a sickly condition, are met with clinging to the grass, or bushes. A careful examination of them will show a number of little bright-red mites crawling on them, or attached near the base of the wings, and evidently the cause of their weakness.

Curious cases have been found wherein these creatures had met with fatal accidents. In their headlong fall to the

FIG. 89.—GRASSHOPPER PIERCED WITH SPEAR OF GRASS.

ground, after one of their reckless jumps, they are liable to have their armor pierced with the dried spears of grass. Fig. 89 represents a grasshopper which had been pierced in this way, the dried point of the grass probably striking the head, and then glancing off, and entering between the head and the thorax.

86. In studying the early stages of the mosquito, it was found that at the outset the animal breathed air through an opening in the hinder part of the body; that soon after this the opening closed, and air was taken in by two openings on the back, but in no case did the insect breathe through its mouth. In the perfect insect, as well as in most larvæ, there are little openings along the sides of the body. These little openings communicate with tubes which branch, and subdivide again and again, sending their little air-twigs into every part of the body, even into the legs and the veins of the wings. These little tubes represent the lungs of an insect. They necessarily render the body very light besides.

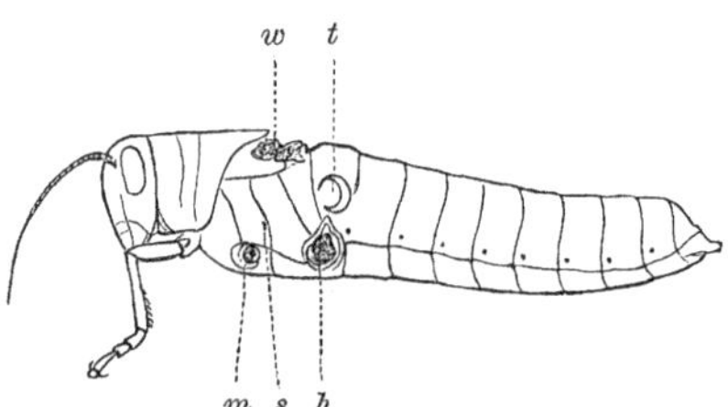

FIG. 90.—INSECT SHOWING THE SPIRACLES.—Grasshopper with the wings and two hinder pairs of legs removed to show spiracles, or openings in the sides of the body which communicate with the air-tubes within the body: *w*, showing where the wings were attached; *h* and *m*, where hind and middle legs were attached; *s*, spiracle on thorax; *t*, tympanum.

In large insects like the grasshopper the minute openings in the sides of the body can be plainly seen without the aid of a glass. The segments of the abdomen have each a little opening, which is represented in the figure; and where the abdomen joins the thorax, a cavity lined with a delicate skin will be found, which is called the *tympanum*, and is supposed to be an organ of hearing. If the softer parts within the body of an insect be removed and slightly compressed between two pieces of thin glass, the air-tubes, looking like fine white threads, may be seen with an ordinary pocket-lens.

The air-tubes are called *tracheæ*, and the openings on the outside of the body which communicate with them are called *spiracles*.

87. Insects breathe by dilating and contracting the abdominal segments. The act of breathing can be plainly seen in the grasshopper or the honey-bee, and it will be noticed that after violent exercise, as in a long flight, the insect breathes more rapidly than when it has been at rest for some time, just as a boy after running finds himself compelled to breathe rapidly for a while.

After violent exercise the insect gets tired and rests. Bees may often be seen, after a long flight, to alight in the grass near a flower, and for a while appear so fatigued that they cannot reach the flower, but remain breathing very rapidly. Insects have curious ways of resting and sleeping. A species of wasp has been observed soundly sleeping while holding on to a blade of grass by its jaws alone, the fore-legs just touching the grass, while the body and the middle and hinder pair of legs were hanging downward, and not bearing against the grass at all, as shown in Fig. 91.

88. In this connection it may be well to allude briefly to

the manner in which the various sounds emitted by insects are made. It is obvious that the vibration of the wings produces the loud buzzing sound made by certain insects. But there are other sounds which are traced directly to the effect of the air rushing in and out of the spiracles, and impinging on certain plates whose sharp edges border the spiracle. The experiment has been made of closing the

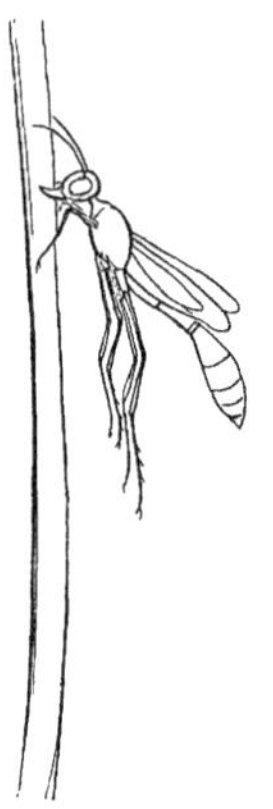

FIG. 91.--WASP SLEEPING WHILE HOLDING ON TO A BLADE OF GRASS WITH ITS JAWS.

spiracles with varnish, when all noise ceased. It is believed that the mosquito produces its remarkable tones in this way. Such noises have always been associated with the vibration of the wings, because the noise seems to be made when the insect is flying, but the cause of this is explained by supposing that the violent muscular action of moving the wings also causes the air to be violently thrown out of the spiracles, and as a proof of this it has been found that cutting off the wings of such an insect, while modifying the sound, by no means prevented it being made; and it is a fact also that no sound is produced by other insects whose wings vibrate with great energy.

89. The peculiar chirp of the cricket is made by the fore-wings being rubbed rapidly against the hind-wings upon which they rest; one of the large veins in the fore-wing being thickened and notched like a file, and the wing itself acting as a resonant body in augmenting the sound. The males, only, make this sound; the females are silent; and if the fore-wing of the female be examined, the vein in question is not thickened, nor is it rough, like a file.

Mr. Samuel H. Scudder has stated that the grasshoppers produce their sound, or *stridulate*, in four different ways: "1. By rubbing the base of one wing-cover upon the other, using for that purpose the veins running through the middle portion of the wing; 2. By a similar method, but using the veins of the inner part of the wing; 3. By rubbing the inner surface of the hind-legs against the outer surface of the wing-covers; and 4. By rubbing together the upper surface of the front edge of the wings and the under surface of the wing-covers. The insects which employ the fourth method stridulate during flight—the others while at rest. To the first group belong the crickets; to the second, the green or long-horned grasshoppers; to the third and fourth, certain kinds of short-horned or jumping grasshoppers."

90. Harris, in describing the third method of stridulation,

says that "their instruments may rather be likened to violins, their hind-legs being the bows, and the projecting edge of the wing-covers the strings," and adds that when a grasshopper begins to play "he bends the shank of one hind-leg beneath the thigh, where it is lodged in a furrow designed to receive it, and then draws the leg briskly up and down several times against the projecting lateral edge and veins of the wing-cover. He does not play both fiddles together, but alternately, for a little time, first upon one, and then on the other, standing meanwhile upon the four anterior legs and the hind-leg which is not employed."

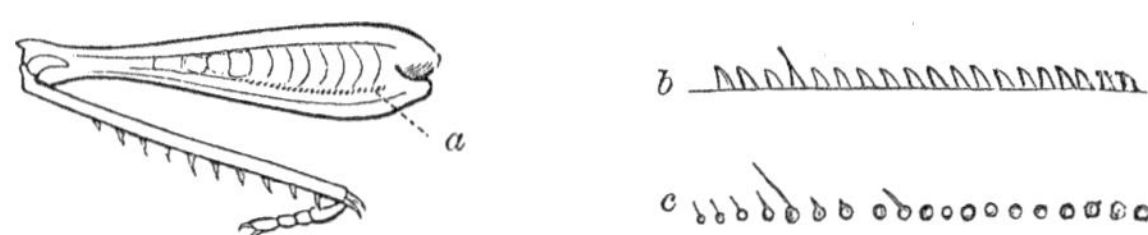

Fig. 92.—Leg of a Grasshopper magnified, showing Ridge of Fine Teeth on the Inside of the Leg, marked *a*, by which the Insect rasps the Wing; *b*, *c*, Different Views of Ridge of Fine Teeth, highly magnified.

A figure is here given of the hind-leg of a common grasshopper, showing the row of minute teeth which occur on the inside of the leg, and which are drawn across the edge of the wing. The pupils may imitate the sound thus produced by drawing a coarse file, or the teeth of a comb, rapidly across the edge of a stiff sheet of paper.

CHAPTER XIII.

HABITS AND STRUCTURE OF INSECTS (CONTINUED).

91. A LITTLE bug, called the tree-hopper, has a peculiar history in its young state. The eggs of certain species are laid in the ground, and, as soon as these hatch, the young ones crawl up the stems of grass, and, piercing the grass with their mouth-parts, commence to suck the juices contained therein. While this action is going on, a clear, watery fluid escapes from certain pores in the body, and in a short time the young insect is completely immersed in it. As it is obliged to breathe air, it secures this by turning up the hinder part of the body, and by means of little appendages, clasping a bubble of air, which then flows along the under side of the abdomen; here it is taken in through the spiracles. The air having been so used, is allowed to escape in the fluid in which the insect is immersed. This operation is repeated over and over again, fresh bubbles of air being thus secured, and then escaping in the fluid. After a while the fluid becomes filled with these little bubbles, which soon convert it into a frothy substance, and this is the origin of the white flecks which occur so thickly on grass, and which is here commonly called *frog-spit*, and, in England, *cuckoo-spittle.*

92. There are certain insects belonging to the same group which are aquatic, and whose young come to the surface of

the water, and in the same manner secure air. So this little tree-hopper, while in the young state sucking the juices of grass, and completely immersed in a watery fluid, may be looked upon as an aquatic larva during this stage.

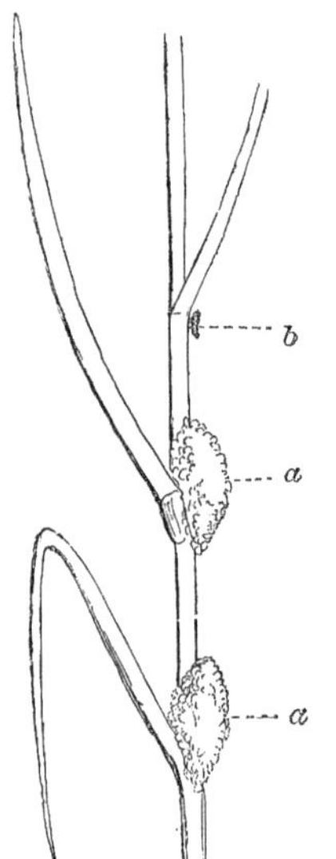

FIG. 93.—GRASS, WITH THE FROTH UPON IT, *a*, *a*, and a Young Insect exposed at *b*.

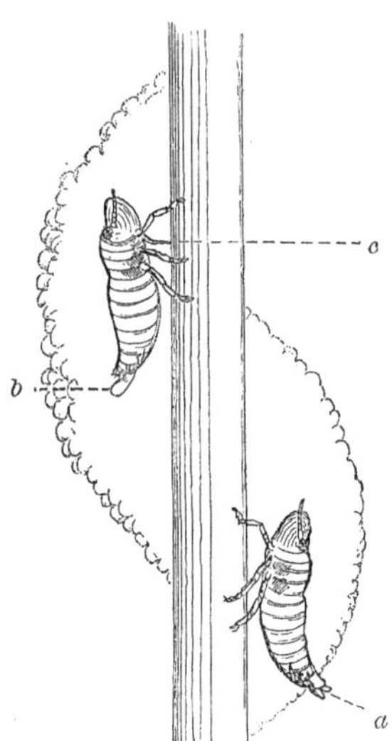

FIG. 94.—A PORTION OF A GRASS-STEM, WITH THE YOUNG INSECTS MAGNIFIED: *a*, the Insect reaching out the Hinder Part of the Body to secure a Bubble of Air; *b*, an Insect allowing a Bubble of Air to escape in the Fluid, the dotted line *b* indicates the bubble; *c*, the Mouth-parts, like a Sting, piercing the Grass.

Let the pupils collect and examine this froth, and, by carefully wiping it away, they may expose the young insect fastened to the grass.

93. The insect called the seventeen-year cicada, or seven-

teen year locust as it is improperly called, has an interesting life-history.

The perfect insect is shown in Fig. 95. They may be known by the peculiar loud, buzzing sound emitted by the male. This sound can oftentimes be heard at a great distance.

FIG. 95.—SEVENTEEN-YEAR CICADA.

The seventeen-year cicada is found rarely in Southern New England, but is common in the Southern and Western States. This species exists in great numbers, and does immense damage to the trees which it infests. The female deposits her eggs in the twigs and smaller branches of oaks. Little furrows are made in the twig, side by side, and into these furrows the eggs are laid. The leaves wither on the trees from the injuries inflicted in this way. Little insects hatch from the eggs, entirely different in appearance from the parent; and these, running to the end of the twig, fling themselves off, and falling to the ground dig their way down, till they come to some root upon which they fasten, and with a piercing sting suck the juices of the root. Here they remain for nearly seventeen years, slowly growing. At the end of that time they assume the appearance represented in Fig. 96. At the proper time they crawl out of the ground, and their skins splitting open along the back, out come the perfect insects, with wings and all complete, to fly away, lay their eggs, and devastate the forests.

There are certain species which do not occur in such numbers, and which pass through all their changes in a single year. The cast-off skins of the pupa of such species may be often found clinging to apple-trees and fences in New England.

FIG. 96.—PUPA-CASE OF A SPECIES OF NEW ENGLAND CICADA OR HARVEST-FLY, CLINGING TO A TWIG.

The pupils should, if possible, collect a twig in which the eggs have been deposited, a pupa-skin, and the perfect insect.

CHAPTER XIV.

HABITS AND STRUCTURE OF INSECTS (CONCLUDED).

94. THE May-fly, or *Ephemera*, is one of the most common insects in the Western States. They live only a few weeks in their perfect state, oftentimes but a few days. Their eggs are laid in the water, and the larvæ live in the water two or three years. At the end of this time they come to the surface in immense numbers, and, shedding their skins, come out as winged insects. At this time they resemble their perfect state so closely, that the name *sub-*

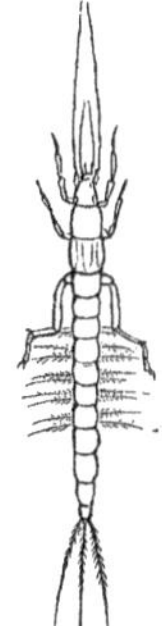

FIG. 97.—LARVA OF EPHEMERA.
(Reduced from Figure by J. H. Emerton, in Dr. A. S. Packard's Directions for collecting Insects, Smithsonian Institution.)

imagos is given to them in this condition. They often fly a considerable distance from the water, alighting on the ground and trees. Here they again shed their skin, and then have attained their perfect state.

These insects occur in prodigious numbers in certain parts of the world. In some regions of Europe they are so

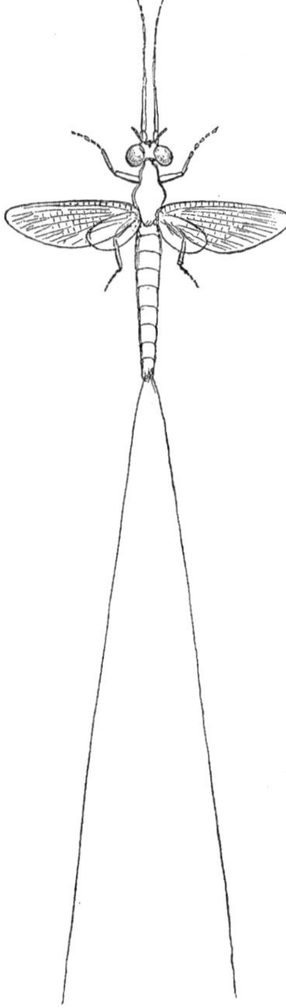

FIG. 98.—EPHEMERA.

abundant that the inhabitants collect them in heaps, and use them as dressing for the land. In the cities bordering the great lakes it is a common sight to see the gas-posts and adjoining buildings blackened by the myriads of Ephemera which have been blown in from the lakes and have been attracted by the lights. The following figure represents a gas-post, in Cleveland, Ohio, as it appeared with Ephemera clinging to it:

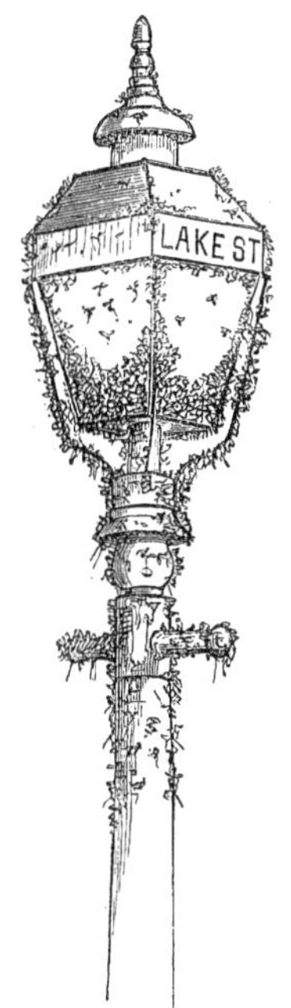

FIG. 99.—GAS-POST, WITH EPHEMERA CLINGING TO IT.

95. Another group of insects somewhat resembling the Ephemera pass their larval state in the water.

Some of their larvæ are called *caddis-worms*, or case-worms, and are inclosed in cases of cylindrical and other shapes. These are variously made of grains of sand, bits of bark and sticks, and other fragments of convenient size cemented together. Some of these cases, built of small

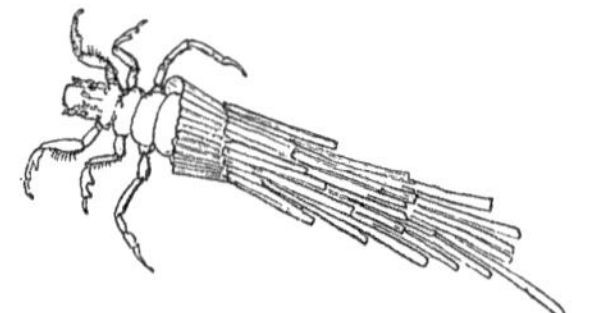

Fig. 100.—Caddis-Worm, with its Case.

grains of sand, look like coiled snail-shells. Other larvæ shelter themselves in bits of straw, or the fragments of hollow stems of plants.

Fig. 100 represents the larva of one of these insects in its case, which is made of bits of sticks arranged in a spiral course. The larva drags about this case, and as it grows collects material for the enlargement of its tube.

In almost any quiet pool or running stream these curious cases may be found, containing the larvæ within.

Fig. 101.—Gall-Fly.

96. The curious round balls called gall-nuts, which are found on the leaves of the oak and of other trees, are produced by an insect called the gall-fly.

The eggs are deposited in the substance of the leaf, and

it may be that the larva, by its presence there, causes the unnatural growth of the leaf, resulting in a wart or tumor, and sometimes in a large round nut. It is believed, however, that the adult insect, in depositing the egg, also stings the leaf, and, poisoning it at the same time, induces the abnormal growth of the leaf. Within this the larva feeds,

FIG. 102.—GALL-NUT ON OAK-LEAF.
(Copied from Harris's "Insects injurious to Vegetation," third edition.)

and changes into the pupa state, and finally into the perfect insect, when it gnaws its way out.

In the autumn the pupils will find the gall-nuts abundantly in the woods. Let them collect a number of these, and, on carefully cutting them open, they will find within a tiny oval case, and upon opening this they will discover snugly stowed away a little, polished black fly having four wings. The creature when liberated is ready to fly away.

Some of the nuts will be empty, because the gall-insects have already escaped.

Galls are also produced by insects not at all like the ichneumon. The following figure, which represents a gall common on the golden-rod, is produced by a two-winged fly. The figure represents the stem or stalk unnaturally swollen, the swollen portion being the gall, within which the larva, pupa, or perfect insect, may be found if the creature has not already escaped.

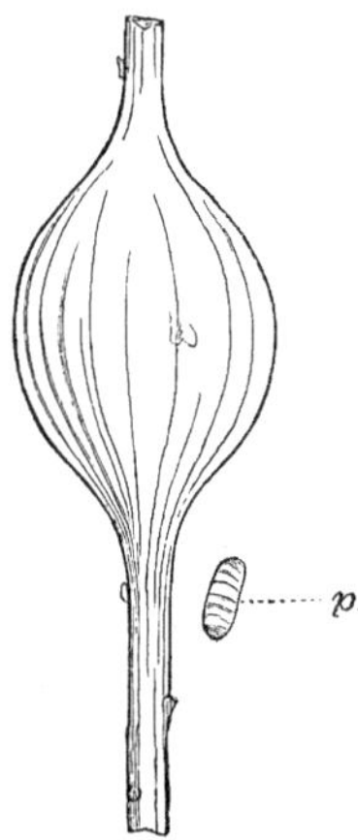

FIG. 103.—GALL ON GOLDEN-ROD STALK.—*p*, Pupa removed from the Gall.

Let the pupils arrange in their collecting-box a leaf with the nut attached, a nut cut open showing the pupa-case, and the insect pinned.

97. Only a few brief lessons have been given representing the life-history of a butterfly, mud-wasp, mosquito, spit-

tle-insect, seventeen-year cicada, May-fly, and gall-fly. Let the pupils endeavor from their own observations to make additional life-histories, or record facts, concerning other insects, such as the honey-bee, paper-wasp, and a great many other common insects, of which no mention has been made here. The turning over of stones and logs in the woods will oftentimes expose the burrows of ants, and the ants will probably be found busily engaged in carrying off long, white, oval cases, which look like eggs; let the pupils collect some of these, and see if they can find out what stage in the history of the insect they represent.

98. An instinct which appears wonderful to us, prompts the insect to seek appropriate places for the deposition of her eggs. The butterfly, for example, seeks for food the nectar of flowers; its larvæ, however, must have leaves upon which to feed, and the instinct of the butterfly impels it to deposit its eggs in a place where the young shall find their appropriate food. It has been learned also that other insects store up animal food for their young, as in the case of the mud-wasp, where spiders are imprisoned in cells in which the eggs have been previously laid.

The gall-flies deposit their eggs directly in the substance of the leaf.

99. Another group of insects, much resembling the gall-flies, deposit their eggs directly in the bodies of the larvæ and pupæ of other insects. They are called *ichneumon-flies*. These insects have on the hinder part of the body a

sharp, piercing sting, and with this organ the necessary hole is made through which the egg is deposited.

A caterpillar soon hatches from the egg thus deposited by the ichneumon-fly, and feeds upon the fatty portions of the body of the larva in which it has been so placed. But this larva containing the ichneumon-caterpillar, meanwhile, completes its growth and changes into a chrysalis, when the inclosed ichneumon-larva devours the entire contents of the chrysalis, and then changing into the pupa state soon emerges as an ichneumon-fly, to go in quest of caterpillars, in which to deposit its eggs. Thus it will often happen that a number of cocoons have been collected, from which ought to appear a certain kind of a butterfly, for example, but from many of them a brown ichneumon-fly will emerge, a sight quite as startling, to one not familiar with insects, as if a robin should be seen to hatch from a hen's-egg.

If the pupils will collect from the fences a large number of the chrysalides of the common yellow cabbage-butterfly, and keep them in a box, with a piece of glass for a cover, they will observe that while a butterfly comes from many, from others, which have already changed to a lighter color, little black flies will appear, crawling out of a hole in the side of the chrysalis which had been made by some of the imprisoned ichneumons. (*See* Fig. 104.)

100. Nearly every species of insect is infested with one or more species of ichneumons, which deposit their eggs either within the pupæ, larvæ, or the eggs themselves.

There are some species of ichneumons which deposit their

eggs within the eggs of the canker-worm moth, and, as tiny as these eggs are, they are still large enough to furnish nourishment and room for the complete development of the insect feeding within.

In Fig. 74 an ichneumon-fly is shown on the wing, in search of caterpillars wherein to deposit her eggs.

Fig. 67 also represents an ichneumon-fly of large size.

Fig. 104 represents ichneumon-flies escaping from the chrysalis of the cabbage-worm moth.

FIG. 104.—CHRYSALIS OF THE CABBAGE-WORM, FROM WHICH ARE SEEN ESCAPING ICHNEUMON-FLIES.

CHAPTER XV.

SPIDERS.

101. FOR this lesson the pupils are to collect a number of spiders, securing, if possible, the largest specimens. A wide-mouthed bottle, with a little alcohol, will answer to collect them in. Let each pupil select the largest specimen to study, and pin it to a piece of cork, or to a soft pine strip. The legs are to be arranged with two pairs pointing forward and two pairs pointing backward, as shown in Fig. 105.

Let them study the following characters with the specimen before them:

The spider is divided into two regions. That region or part to which the legs are attached is called the cephalo-thorax. The hinder region is called the *abdomen*. Instead of having a separate head, as in true insects, the spider has its head and thorax combined, and hence this part is called the cephalo-thorax, a compound word meaning head-thorax.

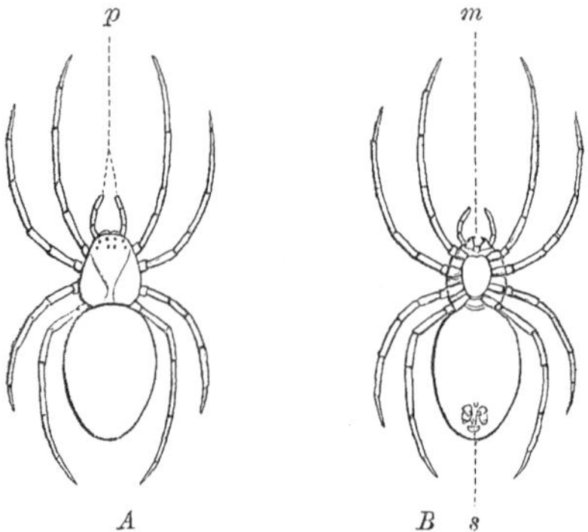

FIG. 105.—COMMON GARDEN SPIDER.—*A*, as seen from above; *B*, as seen from below; *p*, Palpi; *m*, Mandibles; *s*, Spinnerets from which the Spider's Thread issues.

102. The spider has four pairs of legs, instead of three pairs of legs as in the true insects. Projecting in front are a pair of jointed feelers called *palpi* (*see* Fig. 105, *p*). These look very much like legs, and in very young spiders can scarcely be distinguished from them.

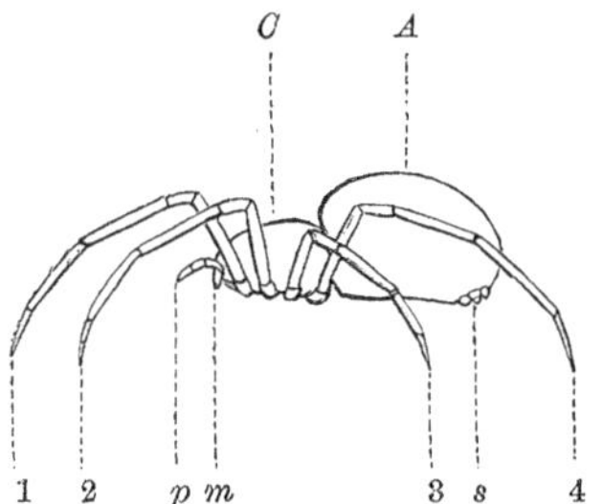

FIG. 106.—SIDE-VIEW OF COMMON GARDEN SPIDER.—*C*, Cephalo-thorax; *A*, Abdomen; 1, 2, 3, 4, First, Second, Third, and Fourth Pairs of Legs; *s*, Spinnerets; *m*, Mandibles; *p*, Palpi.

The mouth is armed with a pair of jaws which are attached above the mouth and hang down in front, at the end of which are the poison-fangs. With these they are enabled to secure and kill the flies and other insects upon which they feed. The following figure represents the jaws or mandibles.

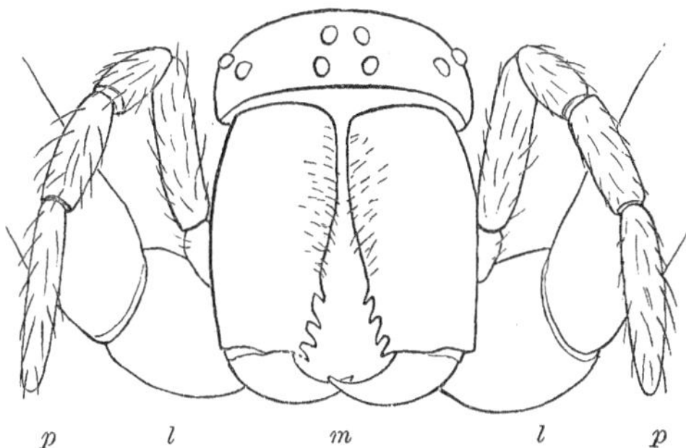

FIG. 107.—FRONT PORTION OF COMMON GARDEN SPIDER GREATLY ENLARGED, showing, *pp*, Palpi; *m*, Mandibles; *l l*, a Portion of the First Pair of Legs, and above, the Front of the Cephalo-thorax shows, with the Eight Eyes upon it.

Directly behind the mandibles, are two smaller jaws, called maxillæ (*see* Fig. 108), which aid in crushing the food and arranging it for the mouth.

FIG. 108.—INNER JAWS, OR MAXILLÆ, OF A COMMON GARDEN SPIDER.—The first Joints of the Palpi are seen also.

The spider has eight eyes, situated on the front part of the cephalo-thorax. They look like little black beads, and in large spiders can be easily seen without the aid of a magnifying-glass.

103. The *abdomen* has little appendages at its hinder end called *spinnerets*, and from these the spider produces the thread with which it builds its nests and nets, the nets being commonly called spiders' webs.

Highly magnified the spinnerets appear as blunt protuberances arranged together in pairs, and capable of being contracted or expanded. These spinnerets are covered with hundreds of jointed hairs which are perforated and through which the web-forming material escapes. This material is fluid and something like the white of an egg. Escaping from the body, through hundreds of these minute openings, the strands of this fluid dry almost instantly, and, uniting, form the delicate, yet comparatively strong, thread of the spider. Thus it will be seen that the thread of the spider is composed of hundreds of strands, which may be often separated just as the fibres of a rope may be pulled apart. Under the microscope the posterior end of the abdomen with the spinnerets looks like this.

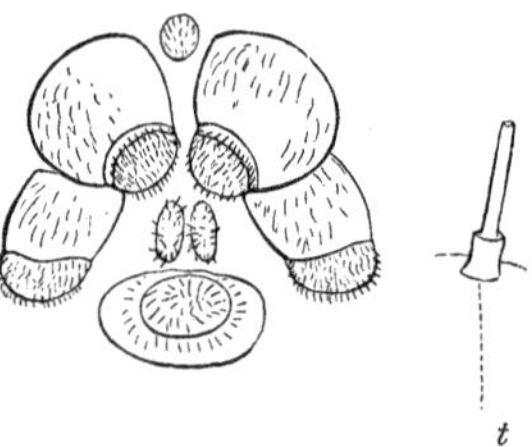

FIG. 109.—SPINNERETS OF A SPIDER.—*t*, one of the Tubular Hairs from the Spinnerets, highly magnified.

104. As the thread issues from the spinnerets, the spider guides it with its hind pair of feet, and these are curiously adapted for the purpose not only of holding and guiding the thread, but also of enabling the spider to run rapidly across its nets without getting entangled, while other animals become helplessly ensnared in attempting the same thing.

The ends of the legs terminate in three claws, a pair of larger ones generally notched like a comb, and a third one like a spine which acts as a thumb. Other notched spines or hairs also aid in securing a hold upon the web, and even if these fail to secure a footing, the leg itself is covered with long stiff bristles pointing downward which are sure to catch in the web. The two large notched claws, as well as the other claw and spines, are highly polished, and consequently present no roughened surface to which the thread will adhere.

The following figure (Fig. 110) represents the end of a spider's leg magnified, showing the arrangement of hooks and claws.

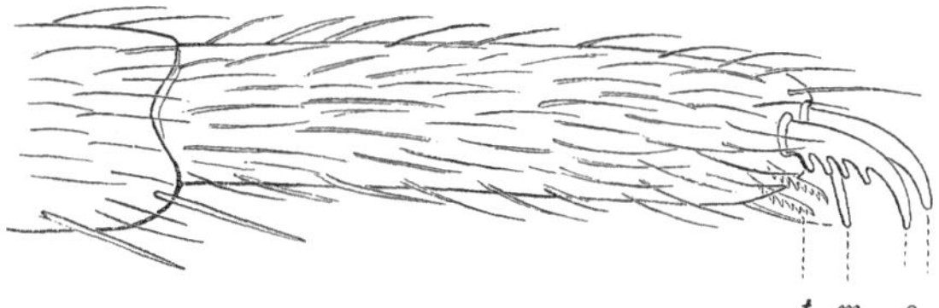

FIG. 110.—END OF A COMMON GARDEN SPIDER'S LEG MAGNIFIED.—*o*, Outer Claws; *m*, Middle Claw; *t*, Toothed Hairs.

105. By observing the spiders which build their nets across the openings of windows and in other convenient

places, and while at work, they may be seen to use their hind-feet in apparently drawing out the thread as it were from the spinnerets. It will be observed that the thread issues in a broad band, and, when these spiders are sluggish, their thread may be caught on the end of a pencil by gently rubbing the spinnerets with it, and then by withdrawing the pencil the thread may be reeled off.

The various kinds of nets are adapted to entrap the spider's food, which consists of flies and other insects.

Certain kinds of spiders do not build nets, but go in search of their prey by stealthily creeping up and pouncing upon it unawares.

It is a very interesting sight to watch the little black-and-white spider (so common on the sides of houses) slyly approach a fly which has alighted near it. If the spider is on the side of a window-sill and a fly has alighted near it, the spider instantly turns round, facing the fly, cautiously and very slowly moves backward, till it gets on the upper side of the window-sill and out of sight, when it rapidly approaches, now and then peering over the edge of the sill, to see where the fly is, and, finally getting directly above the fly, it gathers its legs for a jump, securing its thread to the window-sill at the same time, and then with a sudden spring seizes the fly in its jaws. Sometimes the insect is much larger than the spider, and flies away, with the spider tightly clinging to it; the thread, however, holds fast, though sometimes run out to the length of a foot or more. Soon the poison of the spider takes effect, and the fly gradually weakens, and ceases its struggles, when the spider carries it off to some nook, there to devour it.

106. One of the most common spider-nets is like the one shown in Fig. 73. If the place selected is in the opening of a window or similar place, the spider first runs a few threads as a sort of framework, to which are to be afterward attached the radiating threads, that is, those which run from the centre of the net to the sides. Having arranged these so near together that the spider can easily reach from one radiating thread to the other, the creature commences at the centre of the net, and runs a thread from one radiating thread to the other in a rapidly-unwinding spiral till it reaches the outer edge of the net. This is to form a staging, and also the better to hold the radiating threads in place. It then commences at the outside, and going back over its last course carefully constructs the permanent mesh; and, as it comes to each radiating thread, it will be seen to attach to it the thread it is now making, by simply pressing the spinnerets against it. As it goes around again and again, continually shortening the circle, it gathers up the thread which was first laid as a staging, and, rolling it up in little balls, drops it to the ground. This habit has led to the impression that the spider eats its web. The circular threads are glutinous, while the radiating threads are smooth, and this can be proved by throwing dust through the net, when the cross-threads will catch and hold the dust, while the radiating threads will remain clean. The actual centre of the net is not the geometrical or true centre, but a little above it.

It may be observed, too, that the net does not stand vertical, but leans a little, and the spider having completed the net takes a position in the actual centre of the net, head downward and on the inclining side of the net. With its legs outstretched, and resting on the radiating lines, it can feel the slightest jar or agitation made by a struggling insect. The spider being above the true centre of the net and on the inclining side, if the fly has become entangled below the centre, it can instantly drop to the desired point suspended by the ever-ready thread which it makes, and, swinging to the net, it almost instantly catches the fly.

The pupils would do well to watch the spiders while they are constructing their nets, and to observe and describe, or sketch in outline, the different kinds of nets they find and the kinds of spiders which construct them.

107. Besides the nets made by spiders to ensnare insects, some species have the power of running out a long thread which answers the purpose of a balloon in raising them from the ground and carrying them floating a long distance in the air. In constructing this buoyant means of transportation, the spider does it at peculiar times of the day, and in peculiar positions. Selecting some place where the heated air is rising from the ground or from the side of a fence, it turns up its abdomen and allows the rising current of air to carry upward the thread which is being made, and, when this thread is of sufficient length for its buoyancy to overcome the weight of the spider, it floats away with the spider hanging below.

The following represents the young spider in the attitude of throwing out its thread for the purpose of sailing in the air.

Voyagers often meet with these spiders in myriads as the wind sweeps them from the land.

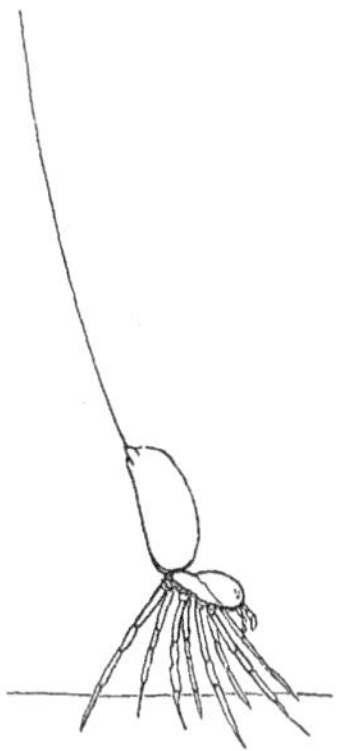

FIG. 111.—YOUNG SPIDER GREATLY ENLARGED, SHOWING ITS ATTITUDE IN THROWING OUT THE THREAD, PREVIOUS TO RISING FROM THE GROUND. (Copied from a Figure by J. H. Emerton.)

108. The spider also constructs cases to hold her eggs, and lines them warmly with the finest web. These nests vary greatly in appearance. A very common variety, somewhat oval in shape, may be found suspended in barns and sheds. The pupils should collect and open these cases or nests, and they will be found to contain little eggs, sometimes rolling out like beads into the hand, or, the eggs having hatched, hundreds of little spiders will appear moving within the nest.

Nests, or, more properly speaking, egg-cases of different kinds, may be collected under stones and logs, and wherever spiders' nests occur. The little spiders hatching from the egg will grow to twice their size in the nest, without apparent food, and it becomes evident that, in some cases, they must eat each other, as Prof. Wilder has observed within some of the egg-cases a far less number of spiders than there were eggs in the nest at the outset. These nests may be kept in boxes, and the eggs will hatch in due time.

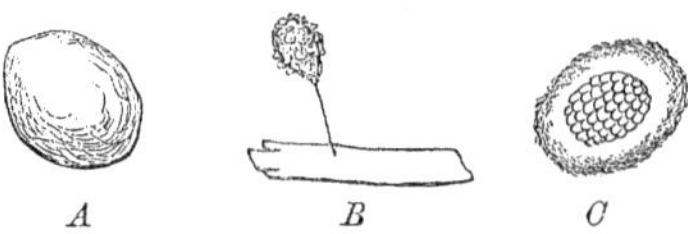

FIG. 112.—SPIDERS' NESTS OF DIFFERENT KINDS CONTAINING EGGS.—*A* and *C* are common nests in sheds and barns; *B* was found under a board in the field, the part containing the eggs stands upon a stalk.

109. The young spider comes from the egg resembling in form the parent spider, except that the legs are much shorter in proportion to his relative size, and the palpi appear so large that they look like another pair of legs, as they then are in fact, but they afterward become modified to feelers.

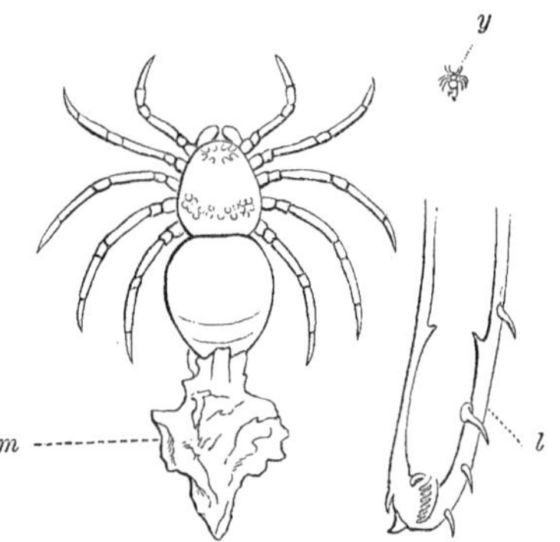

FIG. 113.—ENLARGED FIGURE OF A YOUNG SPIDER JUST FROM THE EGG, WITH THE FIRST MOULT, *m*, adhering to the Hinder Part of the Body; *y*, the Natural Size of the Spider; *l*, extremity of a Leg highly magnified, showing an Outer Skin which has not been shed.

As the young spider grows, it sheds its skin at short intervals of time. If the pupils will examine the young spider soon after it is hatched from the egg, they will find attached to the hinder part of the body the skin which has just been shed. This curious process of shedding the skin, or *moulting*, occurs at intervals, till the spider has reached adult size.

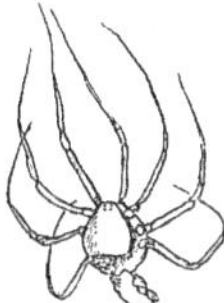

FIG. 114.—THE CAST-OFF SKIN OF AN ADULT SPIDER.

110. The cast-off skins of spiders are very common in their webs, and, if the pupils examine any barn-window which is covered with spider's webs, they will be sure to find some of these cast-off skins, like the one represented in Fig. 114.

The mother-spider, generally so timid, overcomes her fear during the time she has the care of her eggs, and with many spiders the egg-cases are directly cared for by the mother, she oftentimes carrying them about with her or holding on to them and showing the greatest solicitude for their safety. Let the pupils try to separate the egg-case from the mother-spider, and they will then learn how courageous the spider is at this time, and how persistently she remains by her eggs. Some species of spiders carry their young on their backs, and move about with them.

A small black spider was picked up in the woods, which had her body entirely covered with young spiders, which were evidently newly hatched. When the mother-spider was picked up, all the little spiders becoming frightened jumped off, but just before jumping each one attached a tiny thread to its mother's back, and as the spider was held up in the air there hung below, suspended by invisible threads, the whole progeny looking like little black beads. The mother-spider was then thrown down among the dead leaves, sticks, and pine-cones. She did not run away, however, but waited till all of the young ones had found their way through this tangled wilderness, safely back to their mother, and this they accomplished by means of their threads, one end of which they had previously attached to her back. Having waited till all had been gathered in this way, she continued her journey.

111. The spider has no power of throwing or ejecting its thread to distant objects, as many suppose. When threads are seen stretching from one tree to another, the spider has caused the thread to issue from the spinnerets, and the wind has then caught it and borne it along, till finally it gets entangled with some object, and in this way the spider is enabled to cross from one point to another.

These creatures are not so dangerous as many suppose, and but very few authenticated cases are known of man having been bitten by these animals; though the larger spiders at the South, and in California, as the tarantula, for exam ple, can inflict a dangerous wound.

CHAPTER XVI.

DADDY-LONG-LEGS, CENTIPEDES, AND MILLIPEDES.

112. In the insects proper, or true insects, the pupil has learned that the head, thorax, and abdomen, are separated into three regions or parts. In the spiders, it has been seen that the head and thorax are combined, forming a single region or part, and called the cephalo-thorax, while the abdomen appears as a distinct part. There is another group of animals allied to the spiders, the individuals thereof having four pairs of legs, and the head, thorax, and abdomen, more or less merged together. The animals belonging to this group are called in various parts of the country, *daddy-long-legs*, *granddaddy-long-legs*, *grandfather-graybeards*, and *harvest-men*, and in northern New York are known by the name of "grab for gray bears."

Certain species are common around houses and sheds, others are found in the woods. They are easily recognized by their small bodies and extremely long and slender legs. It is difficult to hold them in the fingers, as some of the legs are liable to drop off on the slightest effort made to retain them.

In the middle and on the back of the cephalo-thorax, there is a slight eminence, upon which are situated the eyes, two in number. The abdomen appears distinctly segmented.

113. Their food consists of small insects, such as flies and

mosquitoes; and these they go in quest of, slyly approaching and pouncing on their victim and seizing it with their mandibles, which are furnished at their ends with a pair of nippers, which enable them to retain their prey. (*See* Fig. 115, *m.*)

They build no net to entrap their prey, and are weak and helpless compared with their higher relatives, the true spiders. They are dependent then for food upon such insects as they can overcome, and these they devour, differing in this respect from the rapacious spiders which suck the fluid contents of their prey, rejecting the rest. Certain species are known to

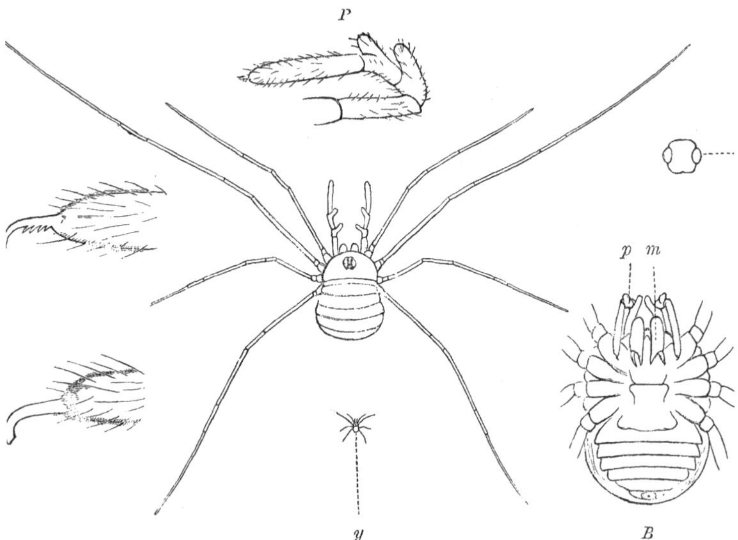

FIG. 115.—YOUNG DADDY-LONG-LEGS, ENLARGED: *y*, showing Natural Size; *B*, Under Side of Body still more enlarged; *m*, Mandible of Left Side; *p*, Palpus of Right Side; *P*, Palpus, greatly enlarged; *P C*, End of Palpus, showing Notched Claw; *L*, End of Leg, showing Claw; *E*, Eye-Prominence, with the Two Eyes. (The three last-mentioned Figures are greatly magnified.)

be cannibals, as some have been seen to pounce upon a brother daddy-long-legs and devour it, leaving only the legs.

It is believed that in the Northern States they do not survive the winter, as in the spring only young ones are seen, and these attain full size by autumn. At this season, the eggs are laid under stones and in the cracks of boards and other protected places, where they remain to hatch out in the following spring.

114. Under old boards in gardens and hidden beneath stones and dead leaves in the fields and woods, the pupils will find the creatures to be studied in this lesson. They are commonly known as centipedes, and in the Eastern States, at least, are also known as earwigs; though the earwig in England is an entirely different animal, being a true six-legged insect.

The centipede belongs to a group of animals called *Myriopods*, and the animals belonging to this group are composed of a great many similar segments, some species having as few as ten segments, others having over two hundred segments. In this latter respect, these creatures resemble the worms, but differ from the worms in having jointed legs and antennæ, and in these last-named characters resemble the insects, besides having other affinities with the insects, in breathing air through spiracles and tracheæ which run through the body.

There are two very distinct groups of Myriopods; one group comprising the true centipedes, in which the body is flattened, the segments loosely joined, and the legs generally equaling, and sometimes exceeding, the width of the body.

The segments in many cases are unequal in length, some of them being very short and alternating with long ones, though all bear a pair of legs below. The antennæ are much longer than the legs, and are often composed of a great many joints. A pair of modified legs reach out behind and look like a hinder pair of antennæ.

In a few forms the eyes are compound as in the insects, while in others the eyes are separate as in the spider, and are called *ocelli*. These are grouped on each side of the head, at the base of the antennæ.

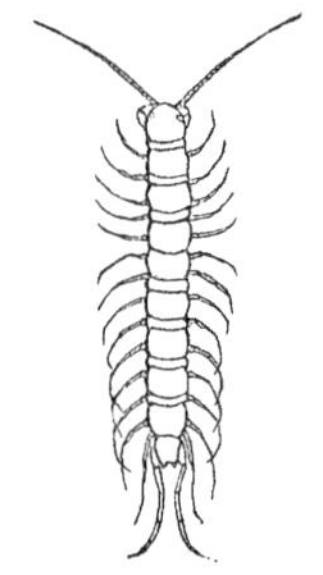

FIG. 116.—COMMON CENTIPEDE, NATURAL SIZE.

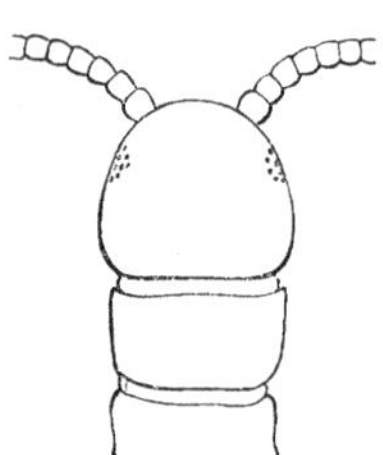

A magnified view of the head, showing group of eyes at the base of antennæ. A few joints only of the antennæ are shown.

115. The jaws or mandibles are large and jointed, with the terminal joint long and sharp as in the spiders. The other pairs of jointed appendages act also as mouth-parts. The under lip is notched with fine teeth, as shown in the following figure, which represents the under surface of the head of the species of centipede shown in Fig. 116.

These creatures are active in their motions, and rapacious in their habits. Some of them feed on small insects, others attack earthworms. Their bite is venomous to insects, and one species having very long legs will produce by its bite a severe pain lasting several hours. A large species found in the Southern States, and in the tropics, and commonly known

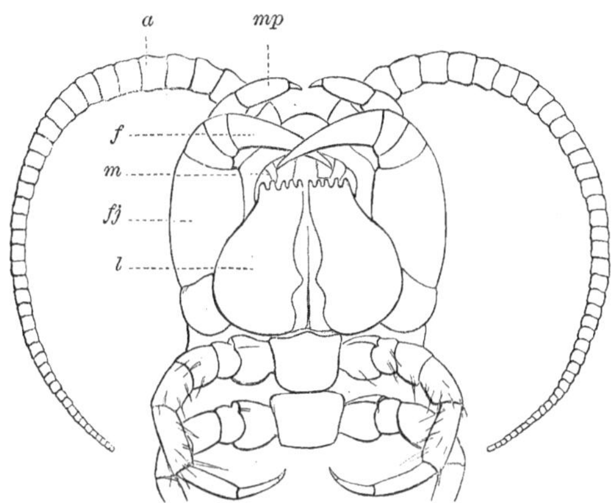

FIG. 117.—SHOWING MAGNIFIED VIEW OF THE UNDER SURFACE OF THE HEAD OF A CENTIPEDE: *a*, Antenna; *m p*, Maxillary Palpus; *fj*, Foot-jaw; *f*, Poison-Fang of Foot-jaw; *l*, Labium; *m*, Maxilla. The Mandibles are hidden behind the other parts, and do not show.

as the centipede, is considered a dangerous animal from its bite. The feet of this species are supposed to poison by their touch, since, when they run over the flesh, small ulcers appear where the feet have come in contact with the skin. The pupils may collect these animals, and either dry them and stick them to cards, or preserve the specimens in vials filled with alcohol.

116. The other group of myriopods, commonly known as

millepeds, have a long, cylindrical, and oftentimes shiny body, composed of a great many segments so smoothly joined together that it is difficult to see the separation between them.

The antennæ are short, there are no long caudal appendages, and the legs are short and feeble. At first sight it would appear that these creatures were exceptional among insects and spiders, in having two pairs of legs to one segment; but it has been learned, by studying the very young milleped, that there is really but one pair of legs to a segment, but that the segments grow together in pairs, so that each apparent segment is composed of two segments.

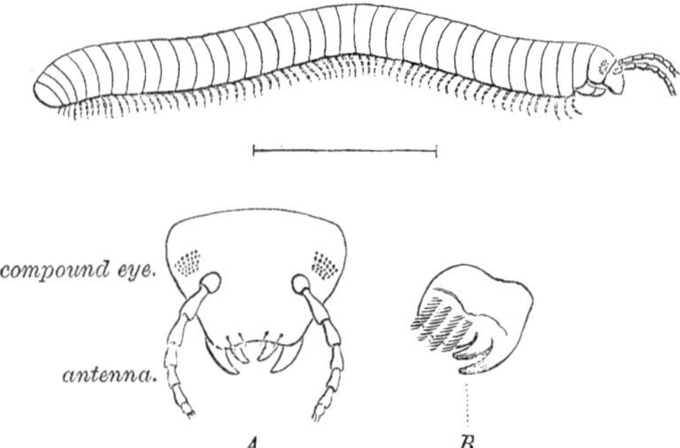

FIG. 118.—A COMMON MILLIPED. The line underneath the figure represents the length of the specimen from which the drawing was made. *A*, a Magnified View of the Head of the Milliped represented above; *B*, a Magnified View of the Left Jaw.

These creatures live on decaying matter, and are slow and weak in all their movements. When touched, or alarmed, they coil up in a closely-wound roll. The body is hard, and the animal can be stuck on a card for the cabinet. The eggs, to the number of sixty or more, are laid in little burrows previously prepared by the creature in earth that is neither too moist nor yet too dry. In preparing the burrow the female makes use of the fluid which comes from her mouth, and which enables her to stick the earth together in little balls, and these she passes up from her burrow by means of the little legs which grasp the pellet and convey it from one pair of legs to the next pair, and so on till it is thrown out of the burrow. After the burrow is completed, and the eggs laid, the entrance to the nest is carefully filled up with clay, or dirt, moistened with fluid from the mouth.

117. It has been learned, in studying the development of the insect proper, that the worm-like larva comes from the egg with its full number of rings or segments, and that, as the creature matures, some of these segments are so merged into other parts, particularly so with some of the caudal ones, that it seemed as if the perfect insect has a less number of rings than the larva. In the myriopods, however, the young creature as it hatches from the egg possesses only a few seg-

FIG. 119.—HIGHLY-MAGNIFIED FIGURE OF A VERY YOUNG MILLIPED, SHORTLY AFTER HATCHING FROM THE EGG.
(Reduced from a figure by Elias Metschnikoff.)

ments, but as it grows new segments are from time to time formed near the hinder part of the body, until the creature attains adult size, when it may possess over a hundred segments.

Like true insects, however, the young myriopod makes its appearance from the egg with three pairs of legs. The body, however, is never divided into a thoracic portion, and an abdominal portion, as in the true insects, or into two regions as in the spiders, but after the head there succeeds a continuous row of similar segments to the tail.

118. In studying the insects, spiders, and centipedes, or myriopods, the pupils have learned something about three groups of animals which have in common a body composed of segments, and possessing jointed legs. They all breathe air through holes in the side of the body, called spiracles, the air, thus breathed, finding its way through various parts of the body by means of little tubes called tracheæ, except in the spiders, where little sacs, called pulmonary sacs, take the place of tracheæ.

In the true insects the segments of the body are gathered into three regions, called respectively the head, thorax, and abdomen. In the spiders the segments of the body are gathered into two regions, called respectively the cephalo-thorax, and abdomen, the head being merged in the thorax. In the myriopods the head is again distinct as in the true insects, but the remaining segments of the body are distinct and are not grouped into regions.

The true insects have three pairs of legs. The spiders have four pairs of legs, while the myriopods have no definite number of legs. In some species there are nearly two hundred pairs of legs, and in no species are there less than ten pairs of legs. The true insects alone have wings.

119. In the growth or development of the true insects and spiders, the young animal comes from the egg with its full number of segments complete, while in the myriopods the young animal comes from the egg with a few segments, and new ones are added as the animal grows.

Some of the characters of the insects, spiders, and myriopods, may be represented as follows:

Three pairs of legs, and having wings.
TRUE INSECT.—Body divided into Three Regions.

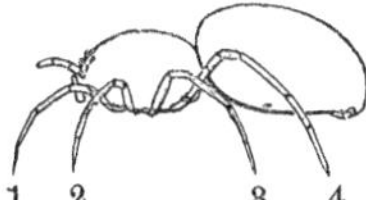

Four pairs of legs.
SPIDER.—Body divided into Two Regions, Head not separate.

No definite number of legs
MYRIOPOD.—Body not divided into Regions, but Head separate.

FIG. 120.—ANIMALS WHOSE BODIES ARE COMPOSED OF SEGMENTS POSSESSING JOINTED LEGS, AND BREATHING AIR THROUGH OPENINGS IN THE SIDES OF THE BODY.

On account of some of these characteristics above mentioned, with others not mentioned, being held in common by the true insects, spiders, and myriopods, these creatures form a natural group in the animal kingdom, just as the snails, mussels, oysters, and clams, possessing certain characters in common, together form a natural group of animals.

There are, however, many other animals which are neither insects, spiders, nor myriopods, and still possess a body composed of segments, and also have jointed legs, and these animals are to furnish the subject for the next lesson.

CHAPTER XVII.

CRAWFISH AND LOBSTER.

120. THE fresh-water Crawfish, or fresh-water Lobster as it is sometimes called, is very common in many of the Western rivers. It may be collected in little pools by the riverside, and kept alive for a long while in a jar of water. It may be fed on the larvæ of insects and fresh-water snails. It would be well to keep the animal alive for a while, so that its motions in swimming and crawling may be observed. For the cabinet, it can be dried with the legs outstretched, or specimens may be preserved in alcohol.

The general form of the body is much like that of the salt-water lobster, differing, however, greatly in size; the crawfish varying from three to five inches in length, and the lobster attaining a much larger size.

The animal is divided into two regions, the body proper, to which the legs and big claws are attached, and the abdomen, consisting of the jointed portion behind. The head does not appear separated from the body as in the insects, but is combined with the thorax, and hence this part is called the cephalo-thorax, as in the spiders. The cephalo-thorax is covered by a continuous shield, or shell, called the *carapace*, while the abdomen is divided into a series of segments. This part can be bent snugly beneath the body (*see* Fig. 121).

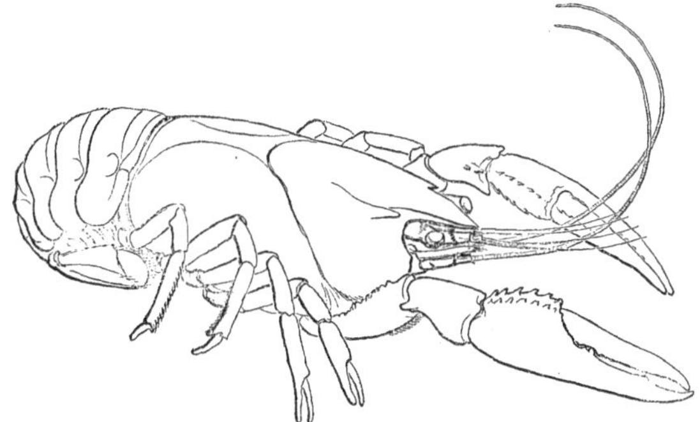

Fig. 121.—Fresh-Water Crawfish from the Mississippi River.

At the hinder end of the abdomen are five flattened appendages, which serve as fins, by means of which the animal can swim vigorously backward. (*See* Fig. 122.)

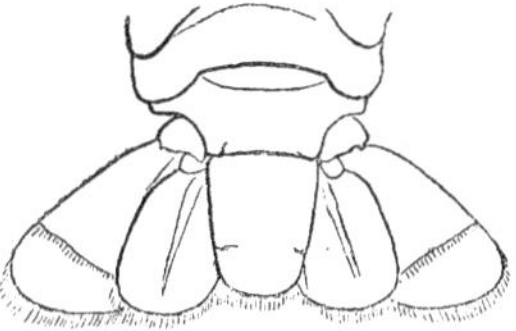

Fig. 122.—Tail of Crawfish showing Flattened Appendages for swimming.

There are two compound eyes in front, which rest upon little jointed stalks, so that the creature can turn them in various directions.

121. Just below the eyes are two long, and two short and double antennæ, or feelers, and directly below these are six pairs of variously shaped and jointed appendages closely packed together. They surround the mouth and assist in securing and preparing the food for the stomach. The first pair are called jaws, or *mandibles*, and are furnished with sharp cutting edges for biting the food, and a flattened surface for grinding or crushing it. The next two pairs are called *maxillæ*, and are accessory jaws. The pair of mandibles and the two pairs of maxillæ, with another pair just behind, making four pairs in all, belong to the head, the other two pairs of mouth-parts belong to the thorax, and are so evidently modified claws or feet that they, with the pair just in front of them, are called foot-jaws or *maxillipedes*. (*See* Fig. 126, in which these parts are all named.)

122. From the under side of the body project five pairs of jointed legs, and these differ in shape and size. The first pair are much larger than the rest, and in the lobster are

Fig. 123.—A Big Claw of the Lobster, showing the Wooden Wedge, *a*.

called the *big claws*. They carry at their extremities big pincer-like jaws capable of giving a sharp nip, and are probably used as weapons of defense, and also to hold on to their prey. The lobster can bite very severely with these big claws, and for this reason the fisherman drives in a little wedge of wood to prevent the animal from opening the movable part, so that he can handle it without being bitten.

The other legs are long and slender. The two forward pairs ending in slender nippers, while the two hinder pairs end in a single projecting claw. With these four smaller pairs of legs the crawfish and lobster crawl or walk.

On the under side of the abdomen are little flattened appendages arranged in pairs, a pair to each ring or segment. The animal not only swims backward by means of the broad

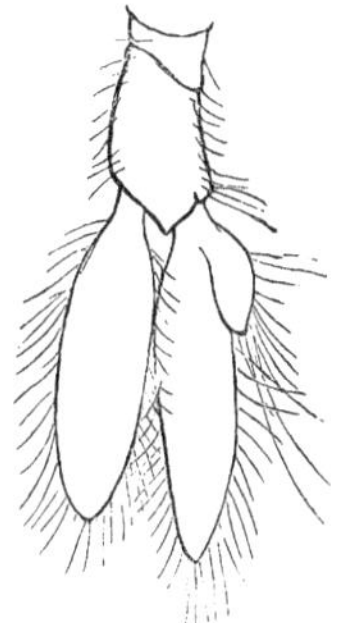

FIG. 124.—ONE OF THE FLATTENED OR ABDOMINAL APPENDAGES OF A LOBSTER.

fins on the end of the tail, or abdomen, but has the power besides to swim in a forward direction by extending the abdomen, and using the little fins below as swimming organs.

123. The crawfish, like the lobster, breathes in the water by means of gills. These are attached to the base of the legs and are concealed on the sides of the thorax by the carapace, which covers them. By forcibly tearing up the side of the carapace, there will be exposed the gills which

look like plumes. This space may be called the gill-chamber, and the water flows into it by passing under the edge of the carapace back of the big claws, and passes out of an opening near the mouth-parts. The currents of water flowing in to the gill-chamber are induced by a stiff appendage attached to the base of the second pair of maxillæ called the *flabellum* (*see* Fig. 125), and which swings back and forth and scoops the water into this chamber. These gills are shown as they appear in the crawfish. In tearing off the claws of the lobster, the gills are often drawn out too, and remain attached to the base of the legs. (In the lobster the carapace can be easily bent up, so as to show the gills.)

In the following figure a crawfish is shown with a portion

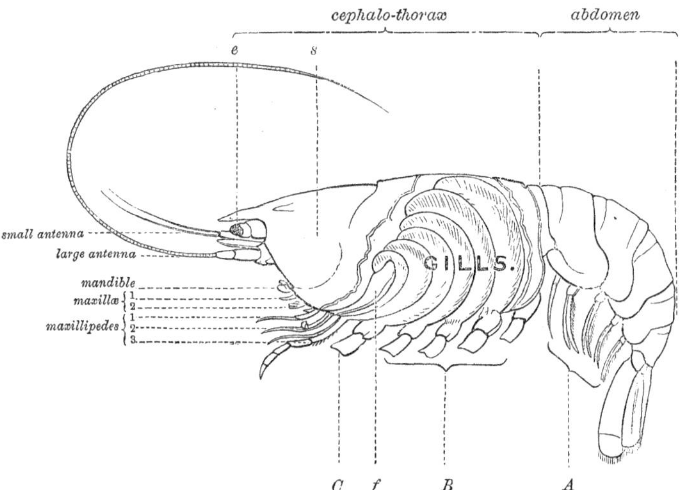

Fig. 125.—Crawfish seen from the Side, with that Portion of the Carapace removed which covers the Branchiæ or Gills. The Appendages of the left side only shown.—*s*, Region of Stomach; *A*, Abdominal Appendages; *B*, Bases of the Four Small Legs; *C*, Base of Large Claw; *f*, Flabellum attached to the Second Maxilliped; *e*, Eye.

of the carapace on the left side removed to show the gills as they appear in the gill-space. The big claw and the four smaller claws or legs are cut off, so that the other parts can be plainly shown.

124. If the pupils are skillful enough, it will be a good exercise for them to separate the various appendages of the crawfish, and arrange them upon a card, just as the beetle was arranged, as shown on page 66, fastening the carapace in the middle of the card with the back uppermost, then gluing the abdomen to the card, or, if possible, separating each ring of the abdomen and gluing each one separately to the card, one behind the other, and then arranging the appendages on each side of the thorax and abdomen, and with a pen marking the names of the various parts on the card.

Figure 126 represents the way in which the mouth-parts of a crawfish or lobster may be arranged.

As the lobster is a much larger animal than the crawfish, it will be easier to separate its appendages, and these may as well be taken from a specimen which has already been boiled, or as it may be obtained in the market. They may then be dried and fastened to a card with glue. The red color of the lobster appears only when the animal is boiled. When alive the color of the creature is a reddish-yellow mottled with bluish or greenish-black.

125. The eggs are carried by the crawfish and lobster glued in masses to the swimming appendages which are attached to the lower surface of the abdomen, and the creatures retain them in this way till the young hatch out.

This feature is characteristic of the class to which these animals belong. How different in this respect from the creatures already studied, in which the eggs are deposited and left by the animal! It has been learned that certain spiders too carry their eggs round with them, and protect them.

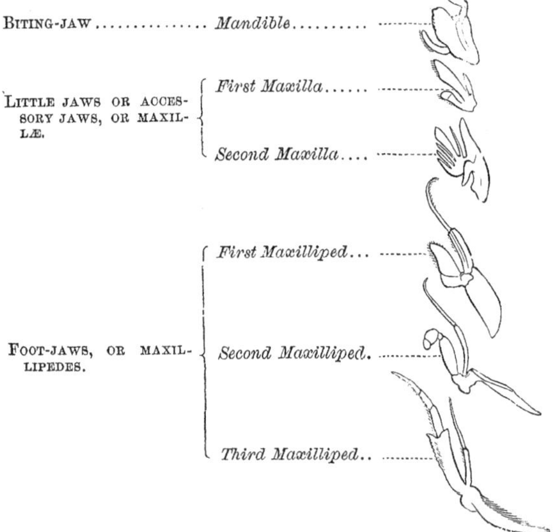

FIG. 126.—MOUTH-PARTS OF A CRAWFISH FROM THE LEFT SIDE.

Fig. 128 represents a crab carrying its eggs glued to the appendages on the under surface of the abdomen. Pupils having access to lobsters in the markets, will, by looking over them, find some specimens in which the eggs are being carried in this way.

126. The young animal in growing sheds its entire shell,

in this respect again resembling the spider. This process is called *moulting*, a term used in describing a similar process in the spider. The lobster and crawfish continue to shed their shells at different periods, till they attain full growth. It is stated that the crawfish sheds its shell annually. If the pupils will keep some of these creatures alive, they will probably have an opportunity to observe this curious process of moulting, in which the entire outer skin, or shell, is discarded, so that there is left, complete in all its parts, the empty crust, like a discarded garment. The carapace separates from the abdomen above and cracks along the back, and by a series of efforts the animal pulls its way out. Great trouble is experienced in withdrawing the legs, and oftentimes a leg is left behind, and cases are recorded wherein the animal has perished in the struggles to liberate itself from the old skin. For some time the animal shows great timidity, and the lobster, when it has freshly shed its skin, retires to some secluded place, and there remains till the soft and tender skin has become thickened and hardened, so as to enable it to withstand the attack of its enemies. Lobsters often lose their legs in fighting, and on a sudden alarm are capable of dropping them off. The loss of the leg in this way is made good by the curious property the stump has of reproducing another leg, which grows out again, jointed and shaped like the one lost, only much smaller than the original one. At each succeeding moult, however, the leg becomes larger and larger. If the pupils will now carefully examine a lot of lobsters, they will notice among them some specimens in which some one of the legs will be much smaller than its mate on the other side. This shows where a new leg has been growing, to replace one previously lost.

127. The young passes through a remarkable series of moults, or shedding of the shell, and each moult brings it nearer in appearance to the general form of a lobster. The following figure presents the appearance of a young lobster which has undergone several such moults.

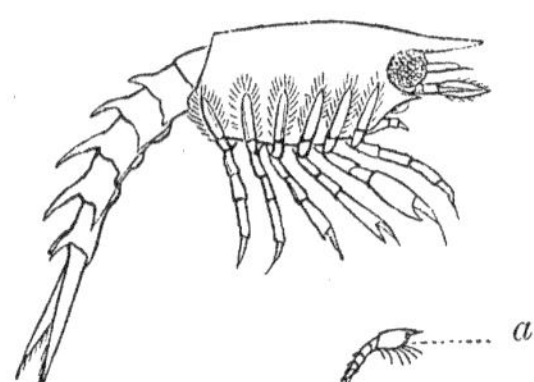

FIG. 127.—A YOUNG LOBSTER MAGNIFIED.—*a* shows the Natural Size of the Creature. (Reduced from a Figure drawn by Sidney I. Smith.)

CHAPTER XVIII.

CRABS, HERMIT-CRABS, AND OTHER CRUSTACEANS.

128. THE class of animals to which the crawfish and lobster belong is called *Crustacea*, a name derived from a Latin word, *crusta*, meaning a crust, or the shell with which the animals of this class are covered.

To this class belong the crabs, hermit-crabs, shrimps, and an infinite variety of forms found in salt-water, certain little creatures found in the great lakes and other fresh waters, as well as a little creature known as the sowbug, which is common under stones and boards in damp places. By far the larger proportion of these animals are found in salt-water.

Pupils having access to the sea-coast will find along the shore, and in pools of water left at low tide, a number of species unlike any thing found in fresh water.

The common crab may be studied and compared with the lobster. It will be found that the crab has the large claws, little legs, mouth-parts, antennæ, and other details similar to the lobster and crawfish. The body, however, is entirely unlike in shape; instead of being long and cylindrical, it is wide and flattened, and the long, jointed abdomen so characteristic of the lobster and crawfish is quite concealed in the crab. At first sight, the creature would appear to have no portion corresponding to this part in the lobster, but beneath the body there will be found a close-fitting piece composed of segments or joints, and, if this be raised or opened, the relation between this small piece and the large jointed abdomen of the lobster becomes at once apparent.

The crab carries her eggs attached in masses to the abdominal appendages which are arranged in pairs on the segments of the abdomen, as in the lobster.

In the following figure, which represents a crab carrying her eggs, a comparison of parts may be made between it and the crawfish or lobster.

The eggs as they are laid are covered with a sticky fluid, which thickens into threads and holds the eggs together and also holds them in masses to the abdominal appendages.

These appendages, having long hairs, retain the eggs all the more securely.

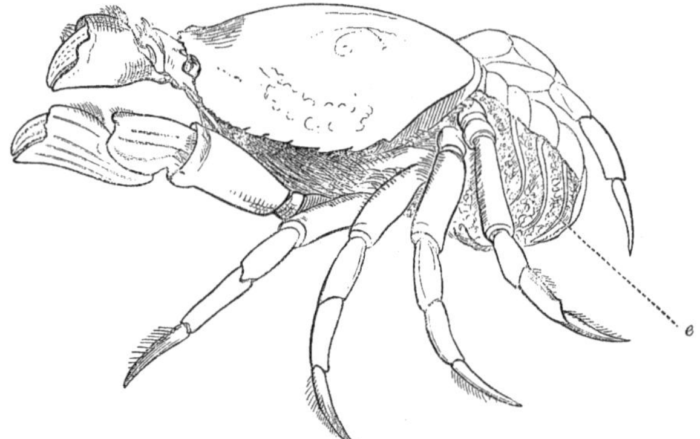

FIG. 128.—SIDE-VIEW OF COMMON CRAB, WITH THE ABDOMEN EXTENDED AND CARRYING A MASS OF EGGS BENEATH.—*e*, Eggs.

Under the microscope the eggs appear like bunches of berries or currants. The following figure represents a few eggs from a common crab:

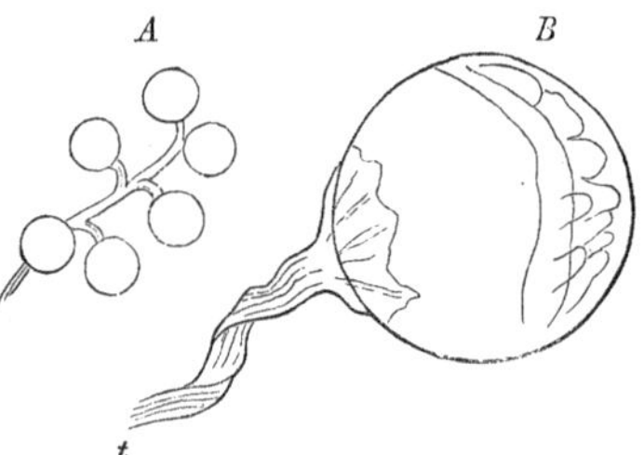

FIG. 129.—*A*, a few Eggs from a Common Crab, enlarged; *B*, Single Egg greatly enlarged, showing more plainly the hardened Thread *t*, by which they are attached to each other. This Egg shows the young crab just beginning to form.

129. The small legs of a crab terminate in a single claw. There are no nipper or pincer-like ends as in the two forward

pairs of claws of the lobster, and in studying the crustaceans generally an infinite variety of modification will be found in these parts. In the crab which is so much sought after for food, and which is known as the soft-shell crab (a condition which indicates that the crab has just moulted, or shed its hard shell), the hinder pair of legs have the last or terminal joints flattened, and these flattened joints are used as fins by means of which the creature swims through the water.

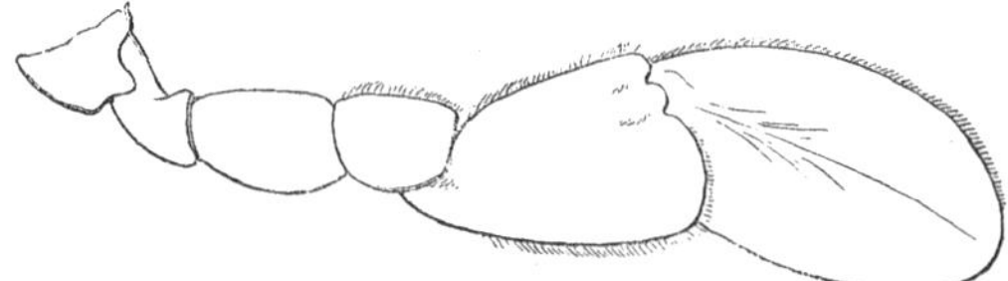

FIG. 130.—RIGHT HIND-LEG OF THE EDIBLE CRAB.

The above figure shows the appearance of the right hind-leg of one of these crabs. Compare this with the common crab shown in Fig. 128.

130. A curious little crab, called the oyster-crab, makes its home within the shell of the oyster, living in the gill-cavity

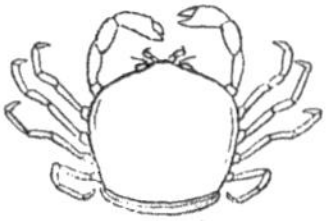

FIG. 131.—OYSTER-CRAB. The Tail is hidden beneath the Body, one Segment only showing.

of the animal. Specimens may sometimes be found in canned oysters, and, to those who do not have access to the sea-shore, these creatures will furnish objects from which an idea of the crabs or short-tailed crustaceans may be got. The tail will be found flattened against the under side of the body. Another species occurs in the salt-water mussel.

In the female crab, a figure of which is given (128), this part is very large and will be oftentimes found holding a mass of eggs. With care the creatures may be dried, and their various parts separated and stuck upon cards for the cabinet.

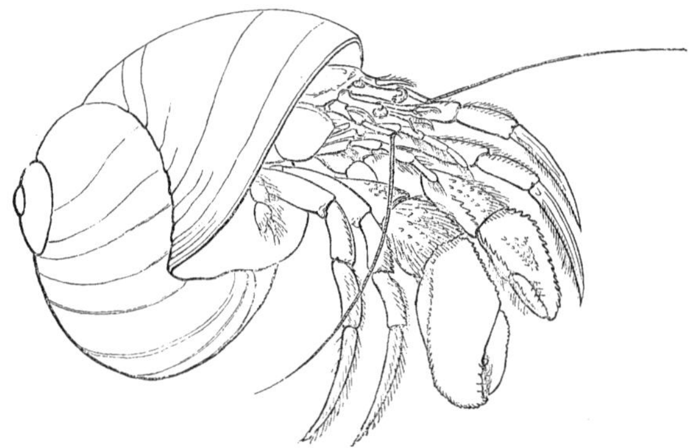

FIG. 132.—HERMIT-CRAB IN THE SHELL OF A SEA-SNAIL.

131. The hermit-crab possesses the general features of the common crab and lobster. The abdominal portion is long and cylindrical, and, instead of being encased in a hardened shell as in the lobster, it is soft and pliant, with scarcely a trace of hardened parts to indicate the segments. The creature, having this defenseless part, protects itself by securing the hard shell of some sea-snail as a house in which it constantly lives. The caudal appendages are curiously modified

to enable it to retain its hold on the shell, and the other abdominal appendages are rudimentary or wanting on the right side, or that side which comes most against the inside of the shell, as if they had been worn off. Wherever the creature goes, it drags the shell after it as a house.

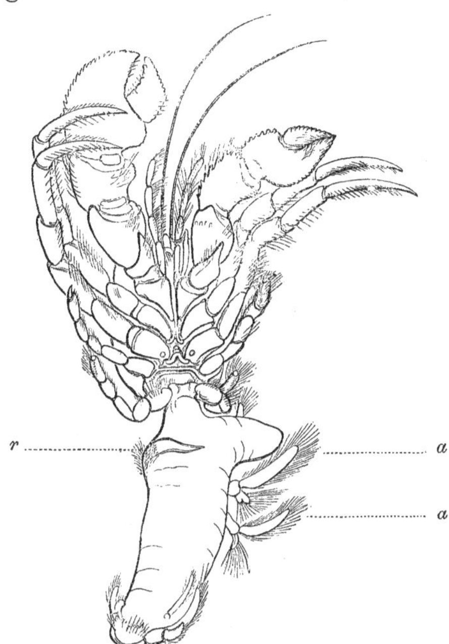

FIG. 133.—HERMIT-CRAB REMOVED FROM ITS SHELL: *r*, Hardened Ridge which bears against the Inner Edge of the Aperture of the Shell; *a*, *a*, Appendages to which the Eggs are attached.

As the hermit-crab grows, it passes through the same features of moulting which characterize the crustaceans generally. The shell which protects its soft defenseless abdomen has, of course, no power of growth, and is abandoned when the hermit-crab gets too big for it. The creature has, therefore, to go in search of another house slightly bigger than the one ready to be discarded. It is said that it does not always content itself with the dead shells that strew the beach, but has been seen attacking a live snail and eating it for the purpose of occupying its vacant shell, and this is rendered probable by the fact that they so often occupy fresh and perfect shells.

Not unfrequently they are found living in old and beach-worn shells which they have dragged about so long as to have worn the shell nearly through at the place where it rests and rubs against the sand. The pupils may collect hermit-crabs of all sizes on the shores.

The figure on the preceding page represents a hermit-crab after its removal from the shell. The creature is drawn as it appears lying on its back.

132. There are comparatively few species of crustaceans found in fresh water; and, with the exception of the species of crawfish and a few others, the fresh-water crustaceans are of small size.

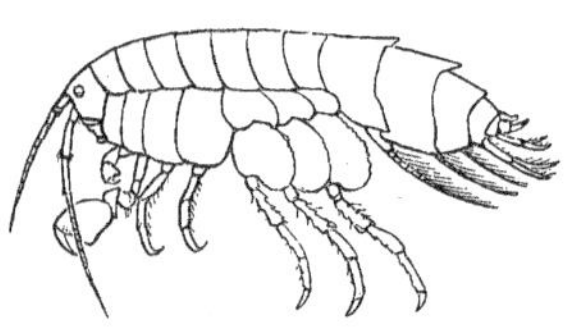

FIG. 134.—FRESH-WATER CRUSTACEAN. The Line below represents the Natural Length of the Animal.

(Reduced from a Figure, in S. I. Smith's Report, of Fresh-Water Crustacea, published by U. S. Fish Commission.)

The figure 134 represents a small species which is common in stagnant pools in nearly all the Northern States.

From this species the pupils may study a form in which the segments of the thorax are not covered by a continuous shield.

The sowbug is a crustacean which lives out of water, though always requiring damp surroundings. It may be collected under logs and stones. In this creature the seven segments of the thorax are easily counted.

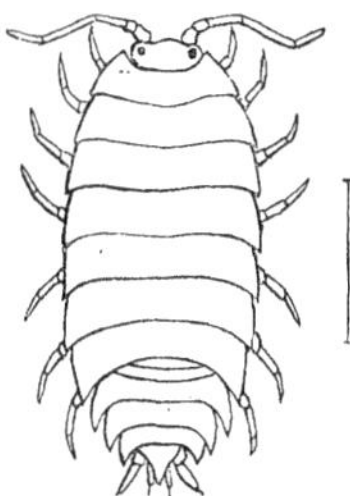

FIG. 135.—COMMON SOWBUG.—The line shows the length of the specimen from which this figure was made.

The eggs of the sowbug, as well as those of other species of crustaceans of the same group, are carried on the under side of the thorax and between the legs, in a little *brooding cavity* made by leaf-like parts which lap over each other and hold the eggs in place.

The eggs of these crustaceans may be found by examining the under side of the body, and observing a lightish-colored space between the legs. With a pin or the point of a knife-blade they may be scraped away without injuring the animal. They are very minute, and only under the

7

microscope can the development of the young creature be watched. The following figure represents a single egg of the sowbug highly magnified:

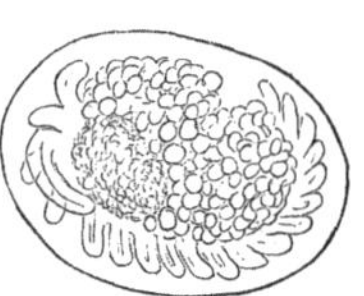

FIG. 136.—EGG OF SOWBUG, HIGHLY MAGNIFIED.—The little dot, at one side, represents the natural size of the egg. The head faces the left.

Around the upper edge of the *embryo* (as a young animal in the egg is called), from eighteen to twenty little blunt appendages may be seen; these represent the legs and other appendages of the body—the one longer than the rest is an antenna. As the creature grows, these appendages become jointed and variously modified to form the legs, mouth-parts, antennæ, and the appendages on the tail, which differ greatly from each other, though at the outset they are all alike.

133. Having studied a few of the many different kinds of crustaceans, let the pupils examine them together to find some points characteristic of them all.

Their bodies, in common with the insects, are composed of segments to which are attached jointed appendages of various kinds. This body is divided into two regions, the cephalo-thorax and the abdomen. In some the cephalo-thorax is covered by a continuous shield, called the carapace, as in the crawfish, crab, lobster, and shrimp. In others the segments of the cephalo-thorax are distinctly separate, and movable upon each other, as in the sowbug and certain other crustaceans, one of which is figured on page 145. The cephalo-thorax is composed of fourteen segments, seven of these belonging to the head, judging from the number of appendages which arise from that part. The abdomen is supposed to possess seven segments, though the last one is so rudimentary that its existence as a true segment has been denied by some.

The deep line running across the back of the carapace, in the crawfish and lobster, is called the *cervical suture*, and represents the dividing line between the head and the thorax.

The following outline represents a sowbug, with the regions of the body marked. Compare this with Fig. 125.

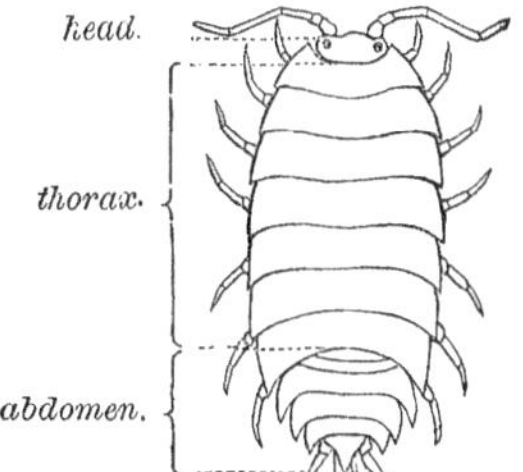

FIG. 137.—SOWBUG, WITH REGIONS OF BODY MARKED.

CHAPTER XIX.

BIVALVE CRUSTACEANS AND BARNACLES.

134. THE crustaceans during their growth shed their hard outer covering at intervals, and, in this as in many other respects, resemble the spiders. In the possession of gills, in-

stead of air-cavities, the crustaceans show a marked difference between themselves and spiders.

Now, there are hundreds of minute crustaceans in which it would be hard to recognize any close relations between them and the crustaceans already studied.

Among these odd forms may be mentioned certain little animals abundant in ditches and pools throughout the country. Some of these creatures are smaller than a pin's-head; others are as large as an ordinary white bean.

At first sight they might readily be mistaken for bivalve mollusks, as the body is covered with a bivalve shell, which partly opens and shuts, and is firmly attached to the body within. If the pupils are fortunate enough to collect some of these creatures in a net and watch them as they actively dart about in a jar of water, they will at once see the difference between them and the clam or mussel.

Instead of the animal's projecting a soft and fleshy foot with which to creep slowly along, as in the mussels, they will see numbers of little jointed, swimming legs partly protruded, and jointed antennæ thrust out in front; and, if their eyes are keen enough, may detect a little black speck just above the ăntennæ, which represents the eye.

The following figures represent a species collected in Dubuque, Iowa, and another form from Lynn, Massachusetts.

135. The concentric lines on the shell appear like lines of growth, and such they really are; but they are not made like the lines of growth on the mussel. When the creature moults, the delicate skin covering the antennæ and swim-

ming legs is discarded. The moulting process also takes place with the bivalve shell, but, instead of its being discarded, the moult is held or cemented to the new shell which forms underneath. Moult after moult of the shell is thus retained, the increasing size of each moult showing as sep-

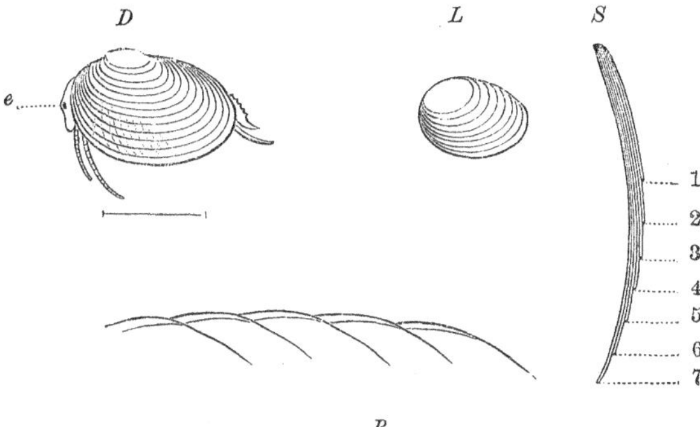

Fig. 138.—Fresh-Water Crustaceans; *D*, Species from Dubuque, Iowa; *e*, Eye. The line below indicates the natural length of the specimen; *L*, Species from Lynn, Mass.: this is figured the natural size; *S* presents a highly-magnified section of one of the shells, to show the successive moults, these being numbered in their order of moulting; *B* shows the appearance of a portion of the edge of the shell along the back, representing the successive moults lapping one over the other.

arate concentric lines of growth. If the shell is cut into and the cut edge is examined with a microscope, the successive moults will be seen resting one upon the other, like the leaves of a book. By reading carefully the description accompanying Fig. 138, the pupils will learn more about it.

The very young stages of these creatures have some resemblance to the young of the barnacle shown in Fig. 141.

136. Another group of animals classed with the crustacea is still more unlike the forms already studied. These are

the barnacles. They are found in immense numbers covering the rocks and piers in places between high and low water mark. In nearly all places along the coast the rocks are whitened by their numbers. Pupils living inland can get specimens of the barnacle by visiting places where oysters are received in the shell; and, by examining the shells as they are thrown away, may now and then come across good specimens. With a stiff brush and some water the mud may be washed off the shell, and then the creature will present the following appearance.

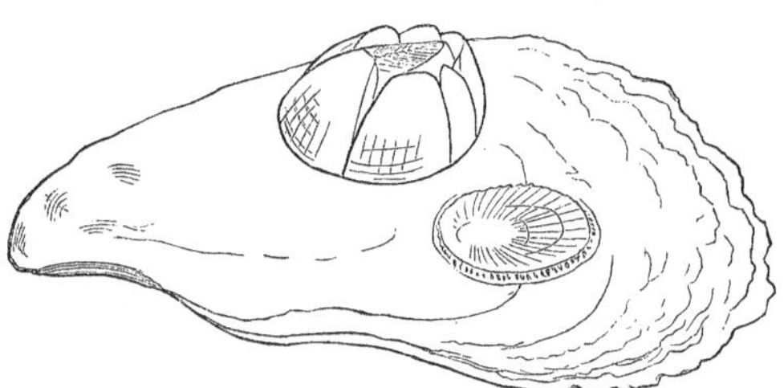

FIG. 139.—BARNACLE ON OYSTER-SHELL.—The circular scar on the shell indicates the place from which another barnacle had been taken.

The barnacle-shell is round and conical, broad at the base by which it is firmly attached to the rock or whatever object it grows upon. The walls of this shell are composed of six pieces whose edges overlap each other. The opening in the top of the shell is closed by four plates which tightly fit together.

In collecting these creatures for the purpose of studying them alive, care must be taken in breaking them from the rock. It will be much better to take them from the wooden piers, or, if possible, specimens should be collected attached to some pebble. These may now be placed in a jar or bowl of salt-water; and, if they are watched closely, there will be seen a set of fringed arms, like hairs, thrust out at the top of the shell, which, stretching out with graceful curves, close and partly retract within the opening. This motion will be constantly and rapidly repeated. There are six pairs of these appendages, and they are flung out in this way to grasp the minute particles in the water, which serve them as food. The arms being jointed and fringed with delicate hairs, the whole combined forms a sort of net. In Fig. 140, *C*, the tip of one of these arms is shown, highly magnified.

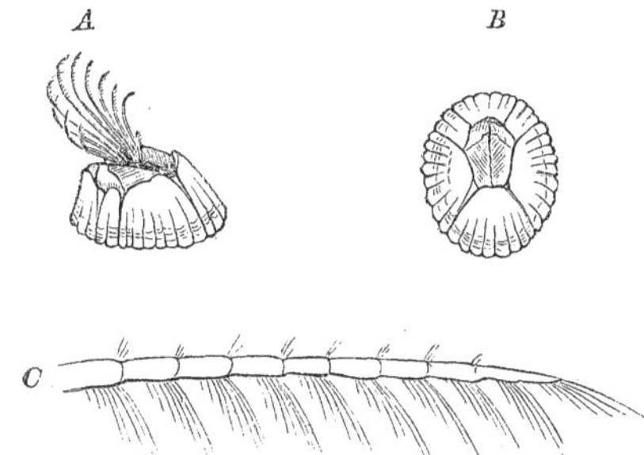

Fig. 140.—*A*, Side-View of Barnacle, Natural Size, showing Appendages protruded for Food; *B*, Top-View of same closed; *C*, highly-magnified View of the Tip of one of the Appendages.

137. Inland pupils may break open the barnacles collected on the oyster-shell, and, taking out the soft parts, may examine them by placing the parts in water, when the fringed arms become apparent; and under the microscope the delicate hairs

which fringe the arms may be seen. The jointed structure of these appendages and their arrangement in pairs show that the creature does not belong to the shell-fish or mollusks, as its shell might seem to indicate.

In past times many able naturalists classed these creatures with the mollusks, because they judged from the external appearances of the shell, which was limy. A careful study of their anatomy and development proved their relations to the crustaceans, and that they had no affinities whatever with the mollusks. In their growth they moult, in this act shedding all the skin, and at certain times in the summer the water will contain myriads of their cast-off skins. The shell, however, is not shed.

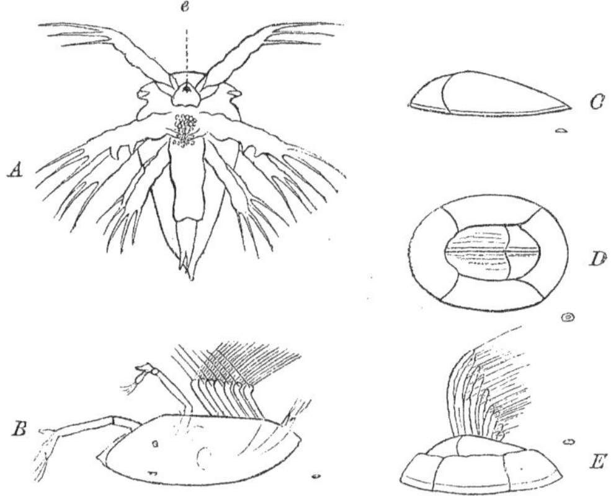

FIG. 141.—EARLY STAGES OF A BARNACLE: *A*, shortly after leaving the Egg; *e*, Eyes; *B*, having acquired a Bivalve Shell, and just before becoming attached, represented upside down; *C*, Appearance after becoming attached—Side-View; *D*, Top-View of still later Stage, with the Shell forming around it; *E*, Side-View of Later Stage, showing Appendages protruded. (The little marks at the sides of the figures indicate the natural size of the object. *A*, *B*, highly magnified; all of these Views are magnified, and, with the exception of *D*, are reduced from figures of C. Spence Bate.)

138. The young come from eggs as free-swimming animals, furnished with eyes and jointed appendages provided with hairs. In this condition they swim about for a while, and then acquire a bivalve shell, and in this state remotely resemble the little creatures shown in Fig. 138.

In Fig. 141, the letter *A* represents its first appearance from the egg. Its next appearance with the bivalve shell is shown at *B*. After remaining a free-swimming animal for a while it becomes attached to the rock, adhering by means of appendages on the head, and, then moulting, it loses its former appearance; the appendages change their proportions, new parts are added, the eyes disappear, a limy shell gradually forms around it, and it assumes characters entirely unlike those of its earlier stages.

CHAPTER XX.

WORMS.

139. Among the animals thus far studied, having a body composed of segments, the pupils have seen that in all cases the appendages were jointed, that is, the legs and antennæ were composed of distinct segments or joints; and, with the exception of the myriopods, or centipedes, the animals possessed a limited number of segments to the body.

In the group of animals now to be studied—the worms—the body has, generally speaking, an indefinite number of segments, and there are no jointed appendages attached to it.

The most accessible worm is the common earthworm. Specimens can be collected by digging for them in damp earth; and they may be found under almost any board or rock which has lain for some time.

The worms may be washed by placing them in a bowl of water, where their movements will soon remove the dirt.

The body is composed of a series of rings or segments, which are alike in form, except those at the extremities, which differ. The body tapers at both ends. The forward or anterior end tapers to a blunt point, while the hinder end becomes broad and flattened.

By carefully watching the worm when it shortens up after a long stretch, there will be seen, projecting from the sides and lower portion of each ring, minute points, which are the ends of little bristles protruding from openings in the sides. These bristles as they move project backward.

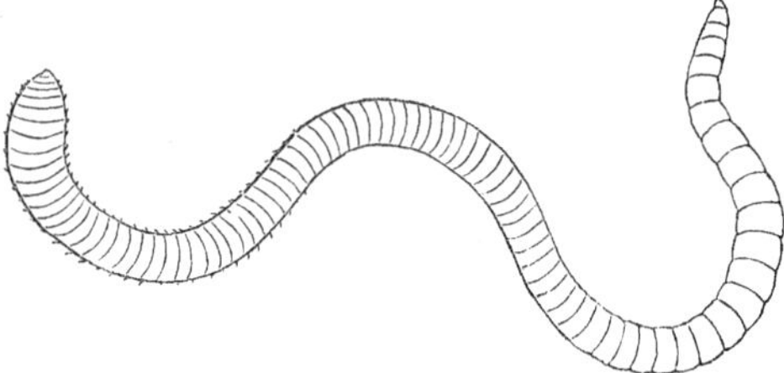

Fig. 142.—Common Earthworm.

The worm moves along the ground by the aid of these bristles, or *setæ*, as they are called. The body alternately lengthens and shortens. When the body lengthens, the setæ on the hinder segments of the body prevent it stretching backward, because the setæ pointing backward stick into the ground; the body therefore can only lengthen in one direction, and that is in a forward one. Then, when the body shortens, the setæ in the forward segments stick into the ground, and the hinder part of the body is drawn up, and by this method the creature moves along.

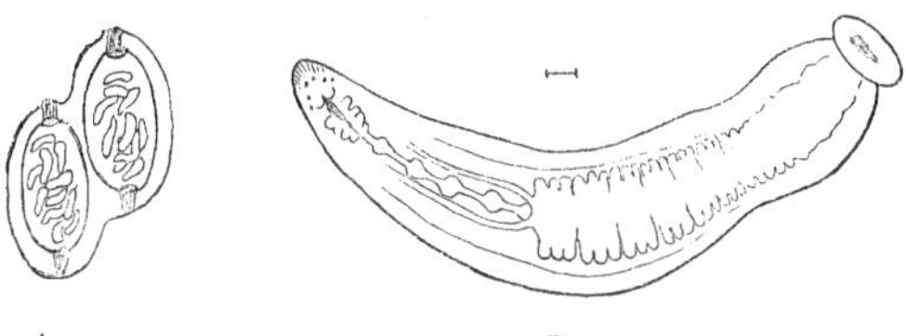

FIG. 143.—EGG-CAPSULES OF LEECH, *A*; AND YOUNG LEECH, *B*.—The egg-capsules are drawn natural size. The young leech, which was taken from the egg-capsule, is drawn greatly enlarged, the little line above showing its natural size.

140. Another very common worm is the leech. Specimens may be collected in almost any pond or lake, and kept alive in jars of water. The creature is flat and broad, and in some species is furnished, at the hinder end of the body, with a sucker, while others have a sucker at each end of the body. It crawls by means of these suckers, and swims through the water by an undulating movement of the body. The eggs of the leech are laid on the leaves and stems of plants which grow in the water. The eggs are contained in little oval and flattened capsules, and these capsules are laid side by side. In Figure 143 *A* represents two capsules, in which the little leeches can be seen; and also a figure of a young leech greatly enlarged, showing the eyes

and mouth, at one end, and the sucker at the hinder end. The body being nearly transparent, the internal organs show through.

The species of worms in fresh water are few in number and quite small. The ocean seems to be their true home, and all along the sea-coast a great variety of worms—many of them of large size—occur.

141. Pupils who have access to the sea-coast may collect them between high and low water mark. Certain species may be got by turning over stones and others by digging either in muddy or sandy places. After a violent storm from the ocean, many kinds of worms are thrown up, and may be found in pools left by the receding tide. The roots of large sea-weeds also afford shelter to certain kinds. All of these creatures may be kept alive for a few days, though considerable care is required, and those not experienced in keeping salt-water aquaria are warned to exclude these animals.

They may be best studied by being placed in shallow bowls or plates, and there will be much to admire in their graceful motions and curious ways.

142. A very common form is found under stones at low tide. The body is composed of a great many segments, from the sides of which project little appendages of various shapes, and also bunches of bristles which can be plainly seen as the creature moves. The head, instead of being simple as in the earthworm, is surmounted by various feelers.

Another species very common on the sea-shore, under stones, is much shorter than the one just described. It has two rows of oval scales along the back, and the mouth is fur-

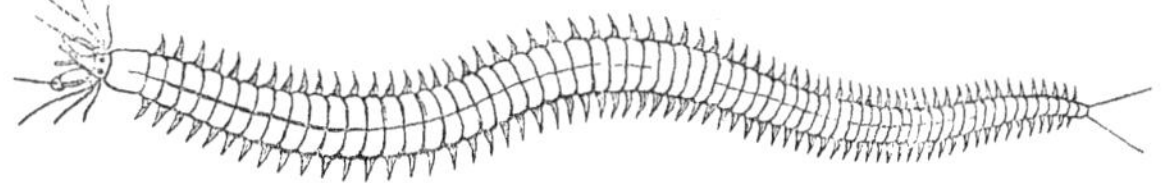

FIG. 144.—A COMMON SEA-WORM.

nished with powerful jaws, which work up and down. When they are placed in alcohol these jaws generally protrude.

143. Some species have a curious way of protruding their œsophagus when they seize their food, at the end of

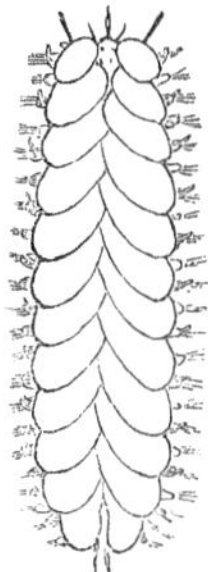

FIG. 145.—A SEA-WORM WITH SCALES.—The eyes may be seen, four in number, between the forward scales.

which appear the jaws, like sharp claw-shaped teeth. The following figure shows the anterior end of one of these worms, with the œsophagus protruded, in the act of securing its prey.

Certain other sea-worms build tubes of mud or sand in which they live, and many of these have bunches of thread-like feelers on the head. If these worms are taken from their tubes and placed in a plate of sea-water in which

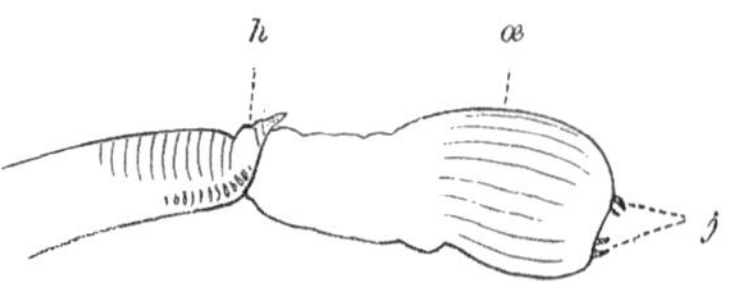

FIG. 146.—ANTERIOR PORTION OF A SEA-WORM, WITH THE ŒSOPHAGUS PROTRUDED.—*œ*, Œsophagus; *j*, Jaws; *h*, Head,

are contained also particles of dirt or sand, they will commence to build a new tube, and for this purpose the threads on the head will stretch out like delicate rubber cords, and, becoming entangled in the particles of dirt, will draw them toward the head, when the appendages on the body will mould it around them in the shape of a tube.

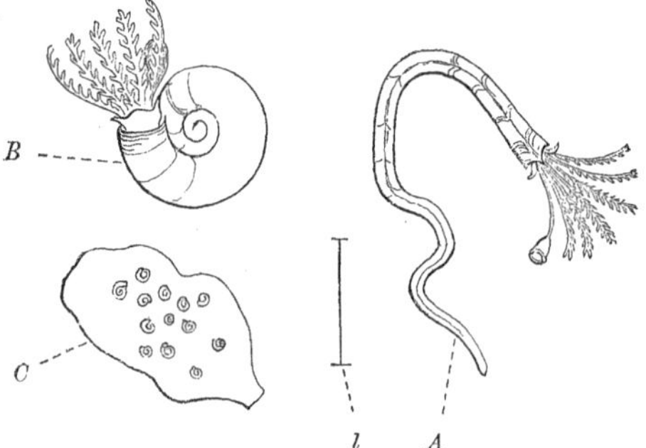

FIG. 147.—SEA-WORMS WHICH MAKE LIMY TUBES.—*A*, Worm with Irregular Tube, the line *l* indicates natural length; *B*, Worm with Spiral Tube, enlarged; *C*, Piece of Sea-Weed, showing the Appearance and Natural Size of these Spiral Tubes attached to it.

Other worms deposit a hard, shelly tube of lime. Some of these are irregular in shape, as in Fig. 147, *A*. Other species build the tube in a coil, looking very much like a snail-shell, as in Fig. 147, *B*. This form is very common on the ordinary rock-weed, and may be collected in the *débris* thrown up by the waves. Pupils should collect these, and, if possible, watch the little creatures as they protrude the feathery appendages which surround the head. In the forms figured, one of the appendages is modified into a sort of plug, and, when the worm retires within the tube, the plug stops up the aperture securely, as certain snails close the aperture of their shells with an operculum.

144. In studying the affinities existing between animals in order to group them together naturally, the pupils should learn how unsafe it is to judge by the external appearances of the animals to be classified. For example, the little worm *B* just described has a coiled, limy shell, which might easily be mistaken for the shell of a snail. Yet the slightest examination of the soft parts within shows that the animal is made up of segments, and that, minute as it is, there are bunches of bristles, or setæ, projecting from the sides of the segments, and from these and other characteristics the creature is proved to be a true worm, having no affinities with the snails. The barnacles too have a limy shell; yet, when the creatures within are examined, their affinities with the crustaceans are seen at once; and, although distinguished naturalists in past times grouped them with the shell-fish, or mollusks, they properly belong to that class which includes the lobster and crab.

CHAPTER XXI.

CONCERNING NATURAL GROUPS.

145. THE pupils have seen, thus far, that the various creatures studied not only differ greatly in their structure, but that some are complex or elaborate in their characters, while others are quite simple. But, while these animals differ so much among themselves, there are certain characteristics which many of them have in common, as in the crustaceans and insects, for example, where all of them had the body divided into transverse segments, and the appendages were all jointed. These features, which are common in large assemblages of animals, are the essential characters by which they are brought together into great groups or divisions. Thus, all those animals which have the body jointed, that is to say, divided into a series of segments, as in the worms, crustaceans, and insects, form the great branch of *Articulates* of Cuvier, because Cuvier, the celebrated French naturalist, first applied the name *Articulata* to an assemblage of animals which included the worms, crustaceans, and insects. Since then other naturalists have had reason to separate the worms from the crustaceans and insects, and to classify them in a great branch called *Vermes*, for worms do not have a continuous hardened and limy skin covering the body and appendages, and, generally speaking, the segments cannot be grouped in regions, as is possible with the crustaceans and insects; the

segments are far more numerous, and there is no definite number of them, as in the two last-mentioned classes; the absence also of jointed appendages and certain features concerning their internal structure, lead some naturalists to recognize, under the name *vermes*, the worms and related forms as a branch separate from the crustaceans and insects, though still having certain features in common with them. The name Articulates is therefore abandoned, and the crustaceans and insects are united in one branch, and called *Arthropods*, a word derived from two Greek words, meaning *jointed foot.* The clams, oysters, mussels, snails, and some other creatures which have not been mentioned in this book, have certain essential features in common, and so they are included in another great branch called *Mollusks*, from a Latin word, *mollis*, meaning soft, because the bodies of these animals are soft. A very inappropriate name, because there are many other animals which are soft bodied having no relation with the shell-fish or Mollusca.

146. Now, these divisions or branches not only include animals which are simple in their structure, but animals which are very elaborate. All the animals in each great division, however, must embrace creatures that possess the same essential characters. With a knowledge of these essential features, it has been customary to make a diagram of a theoretical animal out of these characters only. This theoretical figure is called an *archetype*, meaning an ancient type, or first type, and the characters composing it are hence called *type-characters*, or *typical characters*, and that animal which

possesses most of these characters, in the plainest manner, is called a *typical animal.* This mode of presentation applies as well to smaller groups as to larger ones. Thus in the crustaceans, the lobster and crawfish might be called typical crustaceans, as being the types or representatives of the class, while a barnacle would certainly not be looked upon as a typical animal of this class, though belonging to it. In the same way an insect without wings would not be looked upon as a type of the insects, because one of the leading characters of the class of insects is the possession of wings.

In making systematic tables to show the relative grade an animal occupies, the simplest groups may be placed lowest in the list to indicate their inferior position. For instance, if the arthropods were to be arranged in a systematic table, those which have no lungs, but gills instead, would be placed lowest, because it has been found in other classes of animals that oftentimes the young or immature animal has gills which are afterward replaced by cavities for the purpose of breathing air direct, and the immature animal is regarded as less perfect, or lower in its organization than the mature or adult form. Consequently the crustaceans would be placed lowest in the scale. Then would come the air-breathing arthropods, and lowest among these would come the spiders, as the head is not specialized from the thorax. Next would come the myriopods, as in these the head is specialized as in the insects. And, finally, the true insects would come highest, as here the legs are reduced to three pairs, the head as well as the thorax is definitely separated, and now the creature has added to it wings by means of which it has new means of locomotion. This, however, refers only to the most prominent types.

If the table is to be arranged to show the highest animals in the highest part of the column, it would be arranged as follows:

ARTHROPODS.	AIR-BREATHING ARTHROPODS.	*Insects.* *Myriopods.* *Spiders.*
	WATER-BREATHING ARTHROPODS.	*Crustaceans.*

This is *classification:* to classify animals is to bring those creatures together which have certain leading features in common. And these classes may be divided again and again into smaller groups.

As, for example, the insects proper: if the pupils were to divide them into smaller groups, the beetles would come together as one group, no matter what their shape or size; moths would form another group; the bugs another group, and so on.

CHAPTER XXII.

CHARACTERS OF VERTEBRATES.

147. THE lessons thus far presented have been upon creatures belonging to three great divisions of the animal kingdom, the Mollusks, Arthropods, and Worms. There are many other groups which represent other great divisions of the animal kingdom, of which no mention has yet been made,

and upon some of which a few brief lessons will be given. The material to be collected for this lesson should consist either of salamanders (or lizards, as they are incorrectly termed) or water-newts. The salamanders may be found in groves and forests, under rotten logs or bark. They are absolutely harmless, though many people regard them as poisonous. Water-newts are similar to the salamanders, except that they live in the water, and the tail is often provided with a fin.

The following figure (Fig. 148) represents a species of

FIG. 148.—COMMON YELLOW SPOTTED SALAMANDER.

salamander common to the Northern States, and also found at the South. Its color is bluish black, with a row of irregular shaped yellow spots on each side of the body.

In studying the external characters of the salamander, the pupils will observe the following features:

The animal has a head with a slight narrowing between it and the body. The head has two eyes, capable of being closed by movable lids. The mouth opens transversely; that is, the lower jaw is on the under part of the head, and moves up and down. In the insects and crustacea, the jaws were on the sides of the head and opened sideways. The jaws have minute teeth, and in large salamanders the teeth can be felt by rubbing the finger along the edge of the mouth. On the front of the head there are two holes representing the nostrils. The creature has four short legs, a forward pair and a hinder pair; it has short feet also, with four toes on each forward foot, and five toes on each hinder foot. The tail, which is quite as long as the body at its commencement, is nearly as wide as the body, but tapers gradually, and becomes flattened at its end.

148. How different in every respect is this creature from the animals thus far studied in this book! and yet, if the salamander is compared with a dog or cat, the characters mentioned above will be found in each. The cat has also a head containing two eyes with movable lids; the lower jaw is on the under part of the head and moves up and down and the mouth opens transversely; it is furnished with teeth, there are four legs with feet and toes, and the creature has a tail. But there are also important differences between the two: the cat has external ears, while the salamander has none, though it has parts which enable it to hear. The cat's skin is covered with hair and is dry, while the salamander has no hair upon it and the skin is always moist. The cat has sharp claws,

while the salamander now being described has none. On touching the cat it feels warm, while the salamander feels cold to the touch; with the cat the young are born alive, and the little kittens have the same general features as the parent-cat. The salamander lays a lot of eggs, and most of the species lay their eggs in the water; when these hatch, the creatures coming from them have no lungs and cannot breathe air directly, but have gills instead. More curious still is the fact that, when the young creature hatches from the egg, it has no legs, these appearing afterward as the animal grows.

The following figure represents the appearance of a young salamander ten days after hatching from the egg. The gills, appearing like feathers, are seen on the sides of the neck.

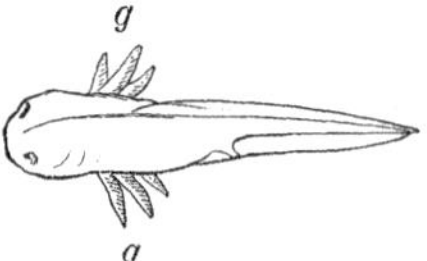

FIG. 149.—YOUNG SALAMANDER.—*g*, *g*, Gills.—(After a figure by Dr. P. R. Hoy.)

149. Thus, while there are important differences between the cat and the salamander, there are also many points of resemblance; and, if the arrangement of bones constituting the skeleton be examined, a still closer resemblance may be seen. Within the body the salamander has a series of bones which together form the *skeleton*. The most important part of the skeleton consists of a row of bones which runs along the central line of the back of the body from the head to the tip of

the tail. This row, or column of bones, is called the *vertebral column*, and the bones composing it are called the *vertebræ*. The bones of the head combined form the skull or *cranium*. The ribs, which in the salamander are rudimentary, are attached to the sides of the body vertebræ. The bones of the fore and hind legs are similar, though they are called by different names.

A series of bones just back of the head, the longest of which is the shoulder-blade, forms the *pelvic girdle;* to this girdle the first bone of the fore-leg is joined. Other bones at the hinder part of the body form the *pelvic arch*, and to this the first bone of the hind-leg is joined.

150. It will be a difficult task for the pupils to remove the flesh from a salamander so as to show the bones united, and it will also be difficult to prepare the skeleton of a cat; but the pupils may learn something about the bones and their attachments by gently handling the creature. Strauss-Durkheim, a celebrated naturalist, when he was writing his famous work on the cat, used to hold one of these animals for hours in his lap, while he felt of the muscles and other portions of the body.

The following figure of a cat shows the position of the bones in outline.

Along the back is a series of prominences which indicates the vertebral column, or, as it is usually called, the backbone or spine. On the sides of the body the ribs may be felt like bars or ridges. The shoulder-blades, or *scapulæ*, are prominent bones, forming the shoulders, and from these the bones of the fore-legs start.

The bones of the pelvic arch, or *pelvis*, may be readily detected, and from these the hind-legs start.

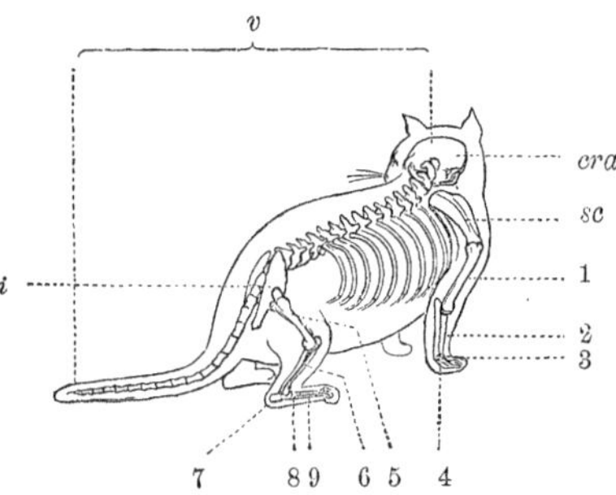

FIG. 150.—CAT, WITH BONES OF RIGHT SIDE DRAWN.—*Cra*, Cranium; *sc*, Scapula or Shoulder-Blade; 1, Humerus; 2, Radius and Ulna; 3, Carpus; 4, Phalanges; 5, Femur; 6, Tibia and Fibula; 7, Tarsus; 8, Metatarsus; 9, Phalanges; *i*, Innominate Bone—a number of Bones combined, forming the Pelvic Arch; *v*, Vertebral Column.

151. This mode of examining the bones is suggested, since it would hardly be possible for pupils to prepare a proper skeleton, and because few of the museums of the country to which they would have access possess skeletons of this kind. Now and then there may be found upon the beach a nicely-cleaned skeleton of a dog or cat, made so by little creatures which have fed upon the flesh, and this may be used for study.

As unlike as the salamander and cat are in some respects, in many characters, both external and internal, they are remarkably alike.

If one of the body vertebra be examined, there will be found a central bony mass, in the upper side of which will be found a hole which is made by the bone growing up from each side of the bony mass, arching over, and uniting above. All of the vertebræ, except those in the tail, possess this channel or tube.

Within the skull is a mass called the brain, and running from this through a hole in the skull is a long white cord called the spinal cord or *cerebro-spinal cord*, and this always runs along on the back of the spine or vertebral column, passing through the arch or hole of each vertebra just described. In fact, the spinal cord is protected from injury by passing through this bony tube.

These features with various modifications will be found in all animals having a vertebral column, that is, a central longitudinal axis either of bone or cartilage above which runs the spinal cord.

The impulses of the animal to move originate in the brain and, passing along the spinal cord, run off by means of nerves, to animate the movements of the muscles. As a proof of this, if the spinal cord be injured, the parts behind and below it are rendered helpless.

The ribs passing from the vertebræ arch below, and form another and much larger cavity, in which are contained the organs which contribute to the body's growth, such as the lungs for breathing, the heart for propelling the blood, the stomach for digesting the food, and so on. The following figure represents a body vertebra to which are attached a pair of ribs with the cavity above, in which is contained the spinal cord or cerebro-spinal cord, and the cavity below, in which are contained the lungs, heart, stomach, etc.

152. If the legs of the salamander are now examined, the following characteristics will be noticed:

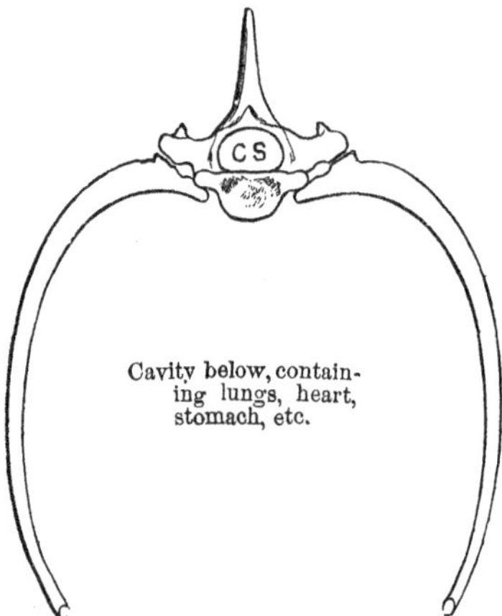

FIG. 151.—A VERTEBRA AND A PAIR OF RIBS, FROM A CAT.—*C S*, Cavity for Cerebro-spinal Cord.

If we compare the hind-leg of the salamander with our own leg, we shall find essentially the same arrangement of parts, namely: the leg bending at a joint in its middle, or the knee-joint, and below this joint another one called the ankle-joint. In that portion of the leg above the knee-joint there is one long bone called the *femur* which joins or articulates with the pelvic arch. In the leg below the knee-joint there are two long bones side by side, called the *tibia* and *fibula*. In the ankle-joint there are a number of small bones closely packed together; these are called the *tarsal bones*, and together form the *tarsus*. In the body of the foot there are several bones upon which the toes rest, and these are called the *metatarsal* bones, and together form the *metatarsus;* and finally come the bones of the toes called *phalanges*—a long series of names to remember, yet they apply to every animal possessing a vertebral column and having legs.

153. The fore-legs of the salamander have similar joints, the elbow-joint corresponding with the knee-joint, the wrist-joint corresponding with the ankle-joint. Above the elbow there is one long bone called the *humerus*, and this articulates with the shoulder-blade, or *scapula*, as it is called. Below the elbow are two long bones side by side, called the *radius* and *ulna*. In the wrist are a number of small bones called *carpal* bones, which together form the *carpus;* and then follow longer bones corresponding with the metatarsal bones, and these are called the *metacarpal bones*, and together form the *metacarpus;* and finally the bones of the fingers which are called phalanges, the same name which applies to the bones of the toes. The following figures represent the bones of the right fore and hind leg of a species of salamander common to the Northeastern and Middle States.

With the aid of a good hand-lens the pupils may see these bones in the leg of any small salamander by observing the following directions: having secured a live salamander, the animal may be killed with ether; now, if the leg be cut off and gently pressed between two thin pieces of glass, the flesh is sufficiently translucent to show all the principal bones quite distinctly:

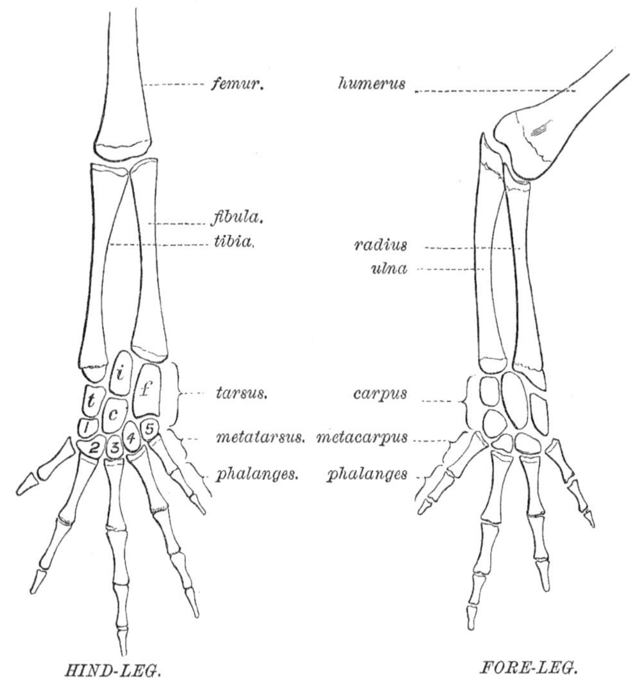

FIG. 152.—HIND AND FORE LEG OF A COMMON WOOD SALAMANDER. (These figures are greatly enlarged.)

154. All animals possessing a vertebral column have essentially the same external features which have been given in the preceding lessons on the salamander and the cat; that is, they have a head containing two eyes with movable lids, a mouth opening transversely and generally furnished with teeth; and, with the exception of the snakes and a few other creatures, possessing two pairs of legs—a fore-pair and a

hind-pair—and internally having a skeleton embracing the features already noticed.

In the fishes the head is continuous with the body, the fore and hind legs are represented by fins, one pair directly back of the head, corresponding with the front-legs, and another pair of fins, which represent the hind-legs, variously situated behind, below, or even in front of the first pair.

155. The animals which possess a vertebral column, and hence called *Vertebrates* (from the Latin word *verto*, "I turn"), are the fishes, frogs, toads, salamanders, snakes, lizards, turtles, crocodiles, birds, and the warm-blooded, four-legged beasts, such as the cat, dog, sheep, horse, and elephant. These last-mentioned creatures belong to a large class of animals called *Mammalia*, the leading features of which are, that the young are born alive, and the mother nurses the young.

The fishes, frogs, toads, salamanders, snakes, lizards, turtles, and crocodiles, and others like them, are cold-blooded, while the birds and mammals are warm-blooded, and all of them, except the mammals, lay eggs from which their young hatch. The fishes, snakes, lizards, and other reptiles are, generally speaking, covered with scales. The frogs, toads, and salamanders, are smooth-skinned; the birds are covered with feathers, while the mammals, with few exceptions, are clothed with fur. In the general grouping of the vertebrates the fishes and amphibians—namely, the toads, frogs, and salamanders—form one group, the reptiles and birds form another group, and the mammals a third group.

With the exception of the fishes, the similarity in the character and arrangement of the bones of the skeleton of every vertebrate is remarkable. Even the birds, which are apparently so different from the mammalia on the one hand and the turtle and salamander on the other, are yet quite similar to each in the general character and arrangement of their bones; and if the young bird, while yet in the egg, is examined, the presence and affinities of certain bones are very clearly seen.

CHAPTER XXIII.

BONES OF THE LEG AND WING OF BIRDS.

156. Our pupils have learned by this time how important it is to study the very young animal in order to determine its relationships. Thus in studying the young barnacle the affinities of the creature were more readily recognized, and in the affinities of the parts of the animal, by studying the young spider the palpi were more easily seen to be modified legs. Now, as an interesting example of the necessity of studying the young or early condition of an animal, a bird is cited.

The fore-leg in the bird is represented by its wing, and in studying the bones of the wing of an adult bird but little resemblance can be seen between them and similar parts in other vertebrates. The humerus, radius, and ulna are plain enough, it is true, and, when the pupil has a chance to pick the flesh from a chicken's or a turkey's wing, he may observe these bones easily enough. The bones of the wrist and hand, however, seem to be few in number and curiously grown together.

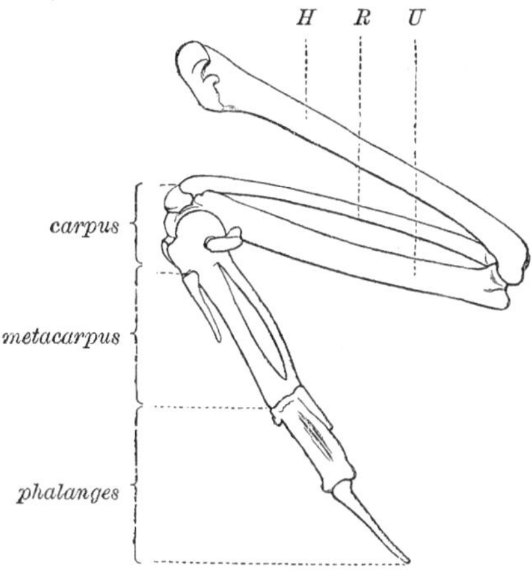

FIG. 153.—RIGHT WING OF AN ADULT BIRD SEEN FROM THE INSIDE.—*H*, Humerus; *R*, Radius; *U*, Ulna.

157. If, however, a young bird is taken from the egg before the parts are fully formed, the bones of the wing will be found separate and distinct, and the relation between the wing of the bird and the fore-leg of other vertebrates becomes more fully apparent.

So constant are the characters of these parts in all birds, that a robin, a sparrow, a pigeon, or a chicken, will reveal the parts quite as distinctly as the larger birds.

In the embryo bird, that is, a bird while still in the egg, the wing and leg appear far more alike than in the adult, as may be seen by looking at the following figures of embryo birds :

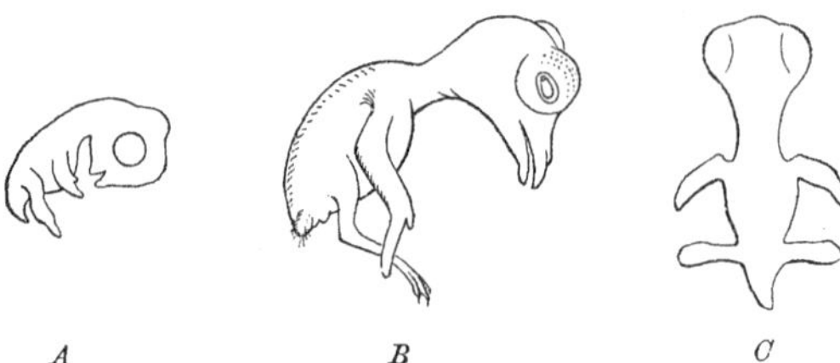

FIG. 154.—EMBRYO BIRDS IN VARIOUS STAGES OF DEVELOPMENT.—*A*, Chipping Sparrow; *B*, Petrel; *C*, Tern seen from above.

If, now, the bones of the wing of an embryo bird be examined, the bones of the extremity of the wing, instead of

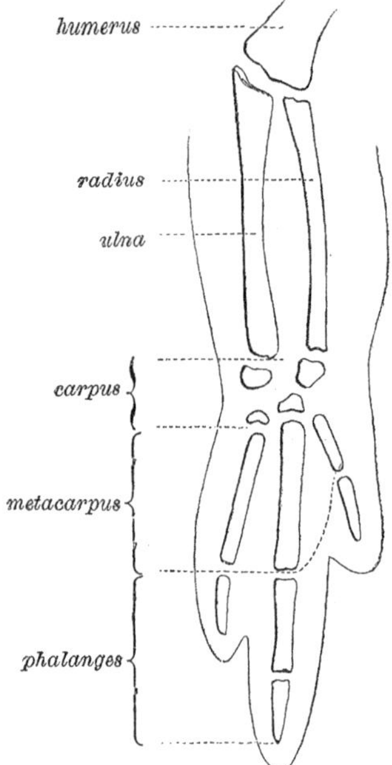

FIG. 155.—RIGHT WING OF AN EMBRYO BIRD GREATLY ENLARGED.—Only a portion of the humerus is seen.

being closely welded together as in Fig. 153, are found to be separate, and the carpal bones, of which only two were apparent in the full-grown wing, are now separate and four in number. The wing at this stage looks more like a three-toed foot. Fig. 155 shows the appearance of the wing of an embryo bird.

As the bird develops, the bones of the fingers gradually approach and some of the bones grow together till they present the appearance shown in Fig. 153.

158. The bones of the leg grow together in the same way.

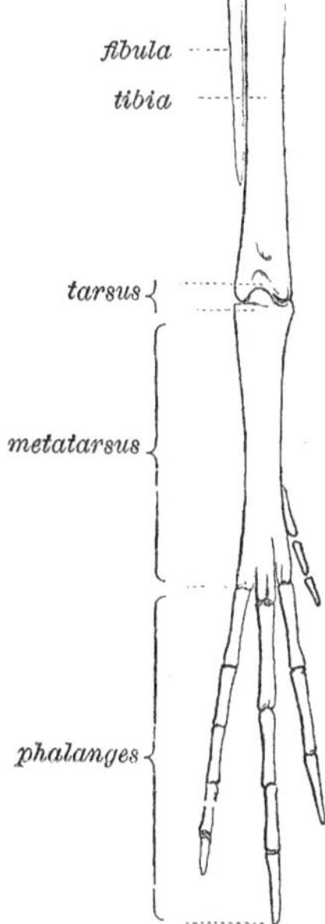

FIG. 156.—BONES OF THE RIGHT LEG OF AN ADULT BIRD.—Only the lower portion of the tibia and fibula is drawn.

The foregoing figure (*see* page 177) represents the lower portion of the leg-bones of a bird.

It will be seen that three of the metatarsal bones, corresponding to the three long toes, are combined; their ends, where the toes join on, appearing separate, while the metatarsal bone of the short toe on the side remains separate from the others.

At the ankle-joint, or tarsus, there are no separate tarsal bones to be seen, nor would their existence be known, except theoretically, without an examination of the embryo.

The following figure represents the appearance of the leg

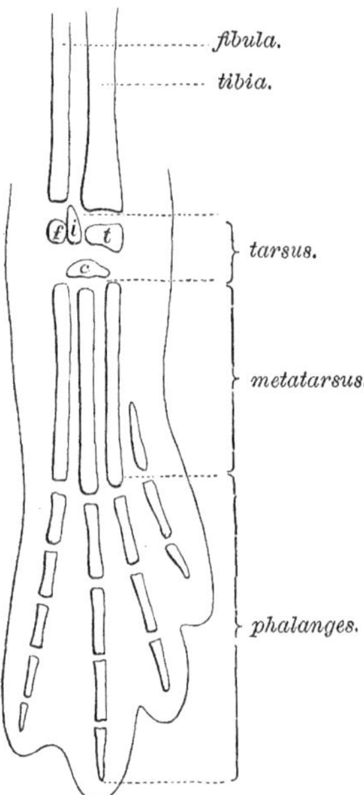

FIG. 157.—RIGHT LEG OF AN EMBRYO BIRD GREATLY ENLARGED.—Only the lower portion of the tibia and fibula is drawn.

of an embryo bird greatly enlarged, showing the bones in place.

At this early stage, not only the metatarsal bones are seen separate, but the bones of the tarsus, four in number, are well marked and distinct.[1]

The many strange modifications in the shape and proportions of the legs of vertebrate animals are accompanied by similar changes in the arrangements, number, and proportion of the bones of these parts.

Thus, in the short leg of the salamander (*see* Fig. 152) the bones of the tarsus occupy a space about as long as that of the metatarsus. In birds, on the contrary, the tarsus is very short, while the metatarsus is very long.

159. If the pupils will observe the hind-leg of the toad or frog, it will not only be seen to be much longer than the fore-leg, but an extra joint seems to be added in the foot. An examination of the bones shows that the first two tarsal bones are very long, while the other tarsal bones are very short; so that in this case the tarsus combined is much longer than the metatarsus. The following figure represents the right hind-leg of a young toad while still in the tadpole state, though the adult shows the same features.

The above brief lessons on the vertebrates do not even embrace an outline of the structure and habits of any of the classes, and they are given only as suggestions toward a method of study, and to point out the essential paths to fol-

[1] The author has seen the tarsal bone, marked *i*, in the heron, tern, and sea-pigeon.

low, in order to gain an insight into the affinities existing between the various groups composing this great branch of animals.

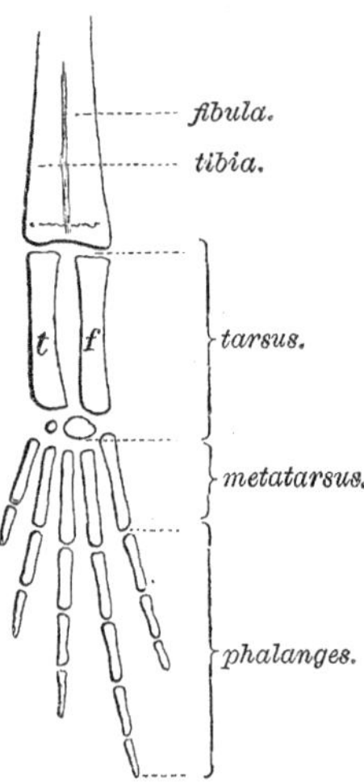

FIG. 158.—BONES OF THE RIGHT LEG OF A YOUNG TOAD GREATLY ENLARGED.—The femur is not shown in this drawing; the tibia and fibula are combined together.

There are still other large classes of animals forming branches or sub-kingdoms quite distinct from those already studied, and represented by animals which live in the sea, such as the star-fish, sea-urchin, jelly-fish, and sea-anemone, of which no mention will be made in this book. There are also many classes, belonging to branches already touched upon, which have not been alluded to. These will be fully dealt with in the second book, now in preparation.

CHAPTER XXIV.

CLASSES AND SUB-KINGDOMS.

THE following figures illustrate the classes and sub-kingdoms which have been dealt with in this book, with their technical names:

SUB-KINGDOM MOLLUSCA (*Clams, Snails, Squids, etc.*).

Class Gasteropoda (snails, periwinkles, limpets, etc.).—Animals whose bodies are generally inclosed in a mantle or sac, which usually secretes a shell composed of one piece, and this often assuming a spiral shape. The body rests upon a broad, creeping disk called the *foot*, and this part represents the ventral portion or belly of the animal. The name *Gasteropoda* is derived from two Greek words, *gaster*, the belly, and *pous*, foot. The following figures represent a few animals of this class:

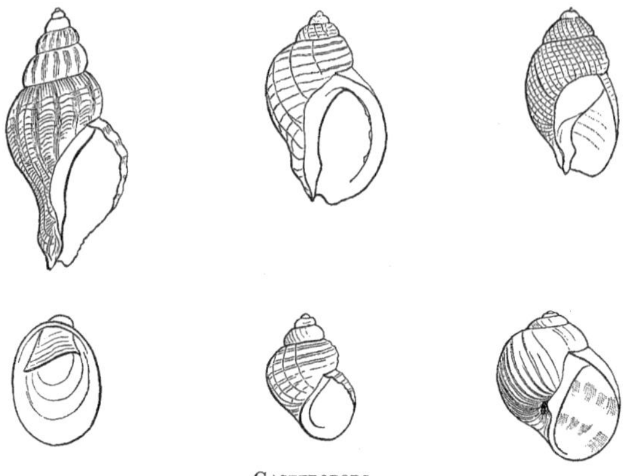

GASTEROPODS.

Class Acephala (clams, oysters, mussels, etc.).—Animals whose bodies are protected by a mantle which secretes a bivalve shell, or a shell composed of two pieces. They have no well-defined head, and hence the name *Acephala*, derived from two Greek words, *a*, without, and *cephale*, head. These animals are also called *Lamellibranchiates*, because the gills form leaf-like membranes or plates on the sides of the body; the word being derived from a Latin and a Greek word, *lamella*, a plate, and *branchia*, gill. The following figures represent a few animals of this class:

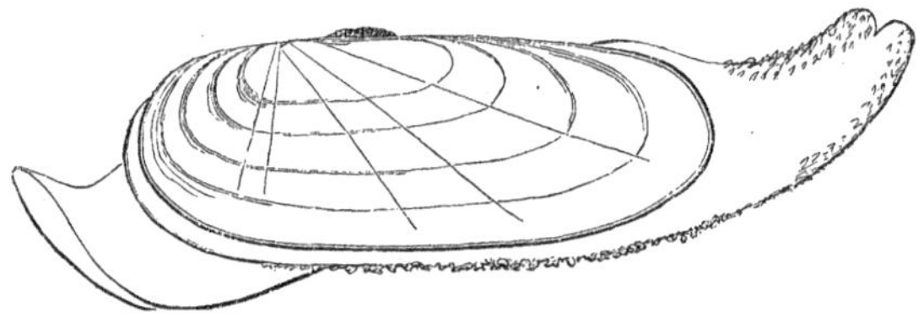

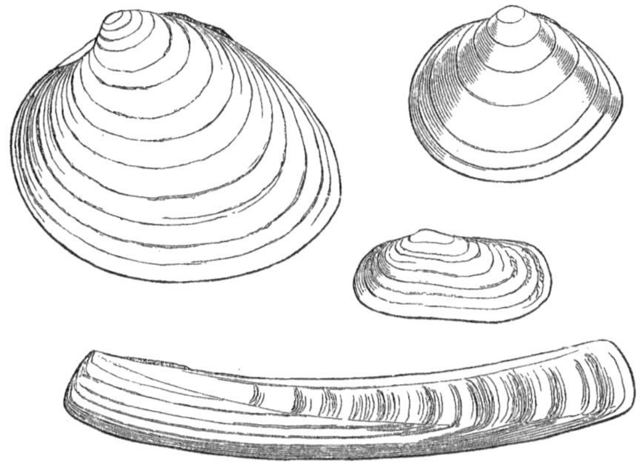

ACEPHALA.

ub-Kingdom Arthropoda (*Insects, Spiders, Centipedes, Crabs, etc.*).

Class Insecta (beetles, bugs, butterflies, etc.).—Animals ıose bodies are made up of segments grouped together in ree regions, the head, thorax, and abdomen; having three irs of jointed legs, and one or two pairs of wings, and eathing air through openings in the sides of the body. ıe word *insecta* comes from a Latin word, *inseco*, I cut into, ferring to the distinct separation of the body into regions.

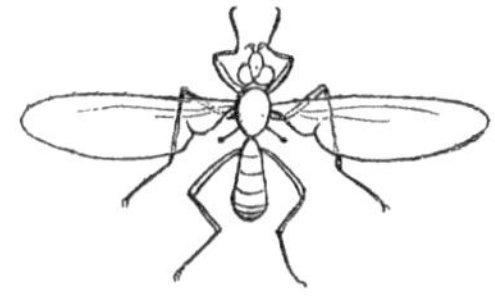

INSECTS.

Class Myriopoda (centipedes, millepedes). Animals composed of many segments. These not apparently combined into regions, except the head, which is distinct. The number of pairs of legs coinciding with the number of segments. Breathing air through openings in the sides of the body.

The word *myriopoda* is derived from two Greek words, *murios*, ten thousand, and *pous*, foot.

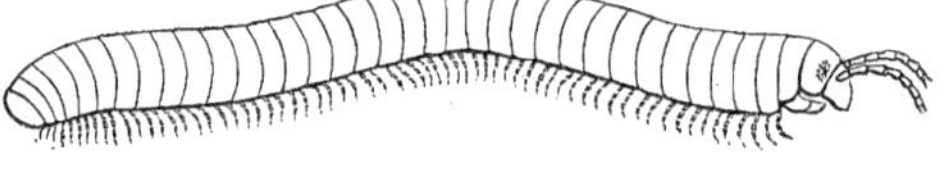

A MYRIOPOD.

Class Arachnida (spiders). Animals whose bodies are segmented. The segments grouped together into two regions, the cephalo-thorax and abdomen. Having four pairs of legs, and breathing air through openings in the body.

The word *arachnida* is derived from a Greek word, *arachne*, spider.

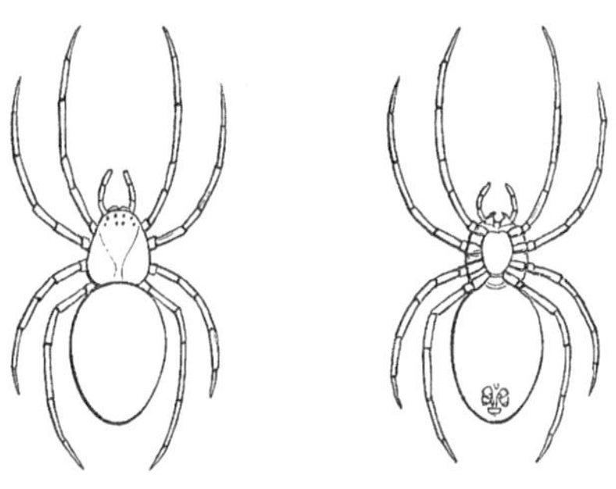

ARACHNIDS.

SUB-KINGDOM VERMES (worms). *Class Annelida* (angle-worms, leeches, certain sea-worms, etc.). Animals whose bodies are made up of an indefinite number of segments, bearing appendages which are not jointed, and in the larger number of groups having bunches of bristles or *setæ* upon the sides of the body which act as supplementary organs of locomotion. The name *annelida* is derived from the Latin word *annulus*, a ring.

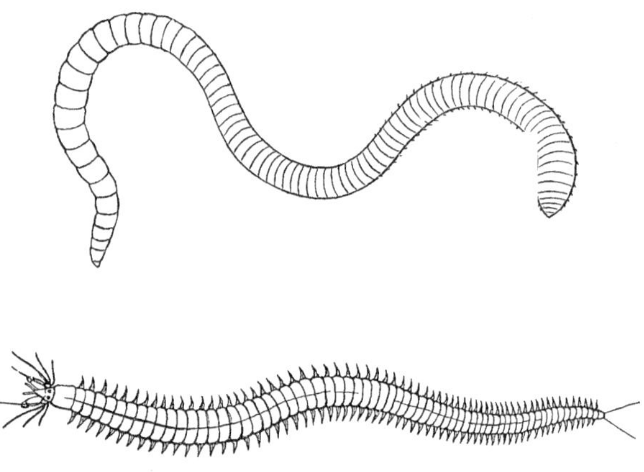

WORMS.

Class Crustacea (lobsters, crabs, barnacles, etc.). Difficu to define, but including animals which pass through a seri of moults in their growth, though in this respect resemblin the spiders, and breathing by means of gills, and in th respect differing from other *arthropods.*

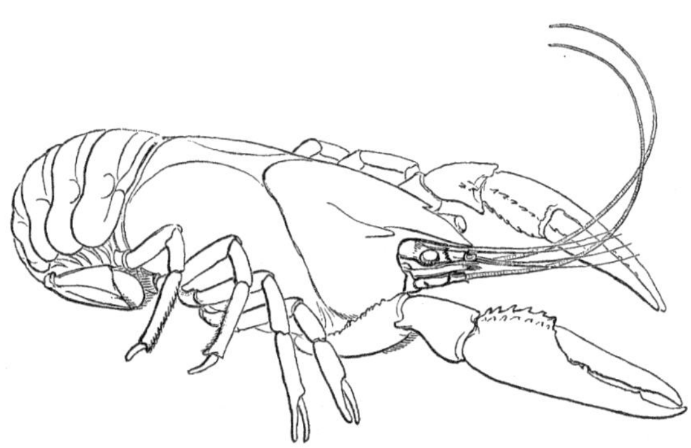

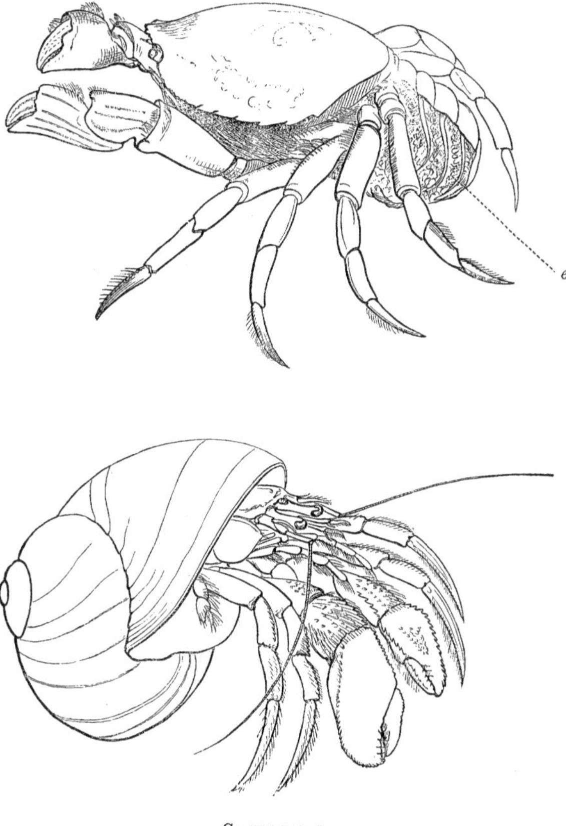

CRUSTACEANS.

SUB-KINGDOM VERTEBRATA, with the following classes, which were only briefly alluded to: *Fishes*, *Amphibians*, *Reptiles*, *Birds*, *Mammals*. According to a late classification of Professor Huxley's, these classes would stand *Ichthyopsida*, which includes the Fishes and Amphibians; *Sau-*

ropsida, which includes the Reptiles, and Birds; and l *Mammalia*.

Ichthyopsida is derived from two Greek words, *ich* a fish, and *opsis*, appearance. *Sauropsida* comes from Greek words, *sauros*, a lizard, and *opsis*, appearance.

VERTEBRATES.

NOTICE TO TEACHERS.

To those who care to pursue the subject more in detail with their classes, or to present the history of those groups of animals of which no mention has been made in this book, the author would suggest the following publications, among many others of value, as works of reference :

Woodward's Manual of Mollusca.
Marine Mammalia and American Whale-Fishery. By Captain C. M. Scammon, U. S. R. M.
Sea-side Studies in Natural History. By Mrs. Agassiz and Alexander Agassiz.
Corals and Coral Islands. By Prof. J. D. Dana.
Packard's Guide to the Study of Insects.
Insects injurious to Vegetation. Harris.
The Annual Reports of the State of Missouri on the Noxious, Beneficial, and other Insects. By Prof. C. V. Riley.
United States Fish Commissioners' Report for 1871–'72. By Prof. Spencer F. Baird. Containing valuable illustrated chapters on the Mollusca, Crustacea, Worms, etc., by Prof. A. E. Verrill and S. I. Smith.
Key to North American Birds. By Dr. Elliot Coues, U. S. A.
Osteology of Mammalia. Flower.
A History of North American Birds. By Baird, Brewer, and Ridgeway.
Canadian Entomologist.
Forms of Animal Life. Rolleston.
Methods of Study in Natural History. By Prof. L. Agassiz.

And for general information, the works of Darwin, Huxley, Owen, Wallace, Mivart, Lubbock, and the Duke of Argyle, and also

Nature, a weekly magazine published in London, and containing an infinite variety of contributions from English and Continental naturalists.

The American Naturalist, a popular illustrated magazine of natural history. Edited by Dr. A. S. Packard and F. W. Putnam, and published in Salem, Massachusetts; containing a vast amount of information by Lockwood, Allen, Coues, Verrill, Smith, Stearns, Scudder, Emerton, Gill, Putnam, Packard, Hyatt, Mann, Marsh, Dall, Cooper, Gill, Cope, Ridgeway, Wood, Abbott, Trippe, Le Conte, Wyman, Dawson, Grote, Mayer, Gentry, Shaler, Wilder, Aikin, Treat, Perkins, Riley, Agassiz, Dana, Hill, Uhler, Edwards, Tuttle, Tenney, Ward, Hagen, Hartt, Shimmer, Hartshorne, Ritchie, Tilsdale, Hoy, Orton, Lewis, Leidy, Brigham, Scammon, Binney, Stimpson, Collins, Fowler, Walker, Jordan, Wright, Norton, Maynard, Canfield, Fellowes, Endicott, and others.

The Popular Science Monthly, edited by Prof. E. L. Youmans, and published by D. Appleton & Co., New York; containing valuable illustrated articles by American and European naturalists.

Every school library should, if possible, contain a complete set of *Nature*, the *Naturalist*, and *The Popular Science Monthly*. For special descriptions of species, the miscellaneous collection of the Smithsonian Institution, with contributions by Stimpson, Gill, Bland, Binney, Prime, Tryon, and others. Also the *American Journal of Science and Art*, and the publications of the Boston Society of Natural History, Philadelphia Academy of Sciences, New York Lyceum of Natural History, Buffalo Academy of Natural Sciences, California Academy of Natural Sciences, Portland Society of Natural History, Museum of Comparative Zoölogy, Peabody Academy of Science, etc.

All these last-named publications should be found in the larger libraries of the country.

THE END.

Lockyer's Astronomy.

AMERICAN EDITION.

"This is by far the clearest and best manual of Astronomy we have ever seen. A child may understand it—and yet it contains information which will be new to all who have not time to follow the latest discoveries." — *New York Daily Times.*

The opinion of the critic of the *Times*, given above, is that of all who have examined this sterling school-book, which is winning golden opinions everywhere. Out of hosts of letters, we have space for the following only:—

"Lockyer's Elements of Astronomy is a work of rare excellence. As a text-book for the use of schools *it is unsurpassed* by any work on that subject with which I am acquainted."—Prof. John S. Hart, *State Normal School, Trenton, N. J.*

"I think it an excellent work—well calculated for class use by pupils of an academic grade. The arrangement and typography are worthy of especial commendation. It is a decided success."—S. B. Howe, *Supt. of Schools, Schenectady, N. Y.*

"It is *the best* school-book on Astronomy that I ever saw. The spectra of the sun, stars, and nebulæ, are worth the price of the book. The diagrams are excellent. I deem it superior to all other books on that science with which I am acquainted. Of course I shall use it."—W. H. Pitt, *A. M., Princ. Friendship Academy.*

"I have examined with much satisfaction the admirable elementary treatise on Astronomy by Lockyer. It furnishes the reader with the means of learning in a short time the great features of the modern progress of Astronomy. No book (except, perhaps, Youmans's New Chemistry) has appeared, which so easily, yet thoroughly, prepares the reader for the subsequent study of that mighty auxiliary to modern science—spectrum-analysis. Moreover it presents, in a clear and succinct style, the relations of the various parts of the stellar universe, besides inducting the learner into that neglected branch of Astronomy known as Geography of the Heavens. Its illustrations are genuine aids to the comprehension of the subject-matter."—David Beattie, *Supt. Troy (N. Y.) City Schools.*

"I have examined it with care, and find it admirably adapted for use in schools. It is so plainly written and so fully illustrated as to render it specially suited for beginners in the science, and, at the same time, profitable for advanced students."—Prof. C. Staley, *Union College.*

"It is a clear and beautiful unfolding of a profound and fascinating subject. It seems to me admirably adapted to the academic grade of students. Not attempting to discuss those problems and theories of the science for which such pupils have neither the time nor the capacity, the author has given an outline of the subject that is clear, sufficiently complete, and thoroughly modern."—Prof. Bradley, *Princ. Albany Free Academy.*

THE FIRST BOOK OF BOTANY

Designed to cultivate the Observing Powers of Children.

BY ELIZA A. YOUMANS.

Price, $1.

This book is incontestably superior to all others of its kind, in two essential respects:

FIRST, it is the best book for *all beginners* in botany, whatever their age, because the rudiments of the subject are reduced to their utmost simplification, and all the study is upon plants themselves. To study botany without studying plants is a mockery of education, because it leads not to real knowledge, but to sham knowledge. Miss Youmans's book involves the study of the plants from the beginning to the end; the book being only a guide and a help.

SECOND, this First Book of Botany is the only school-book we have, that provides systematically for the cultivation of the observing powers of children. It practises them in noticing, searching, comparing, and reasoning, with reference to objects themselves—the most important and neglected part of our mental culture. In short, the plan of this book combines the practical acquisition of one of the most fascinating and useful of the sciences, with that kind of intellectual training which it should be the highest purpose of education to confer.

From Prof. JOHN S. HART, *Principal of the Trenton Normal School.*

This little book seems to me to supply the very thing needed in our course of primary instruction. I refer not to the study of botany, but to the method of studying it here developed. We have here a fourth fundamental branch of study, which shall afford a systematic training of the observing powers; and its general introduction into our primary schools would work a great and most important revolution in our whole system of education.

From the Nation.

We must award the unpretending work of Miss Youmans high praise. The authoress has unquestionably the true conception of the duty of the teacher. Every effort is made to keep the attention of the student upon the object to be studied, and so well has she succeeded that one may safely say that the student can do nothing with the book unless he has the specimen in hand. The plan is so arranged as to be suitable for a primary school, but the method is one which may apply to the college as well. We heartily recommend every teacher in any department of natural science, who is wise enough to doubt the perfection of his methods, to look over this book.

From EDWARD SMITH, *Superintendent of School, Syracuse, N. Y.*

Miss Youmans's Botany is the only work I have ever seen that meets the wants of our schools in the lower grades. I believe it will do more to turn the attention of instructors into the proper channel for the education of children than any thing heretofore published.

From Prof. WM. F. PHELPS, *Principal of the Normal School, Winona, Minn.*

I am delighted with this little work. It gives us a *scientific plan* for the development of the observing powers of the young. Send us at once 150 copies.

After using it several weeks, Prof. Phelps thus closes an elaborate notice of the book in the Winona *Republican:*

Every class in the institution is now devoting a regular portion of its time, daily, to this study pursued in this practical way. The results thus far are highly satisfactory. *It has awakened a new interest in study throughout the school.* Many, who have heretofore been indifferent in their work, have taken hold with great zeal, and are pursuing this fascinating branch with ardor and enthusiasm. It is no uncommon spectacle to see children occupying their play-hours with a bunch of plants, and, book in hand, pursuing this study as a pastime. This, surely, is an unerring test of its value, as it is the highest recommendation that could be given of its adaptation to the wants of our primary schools.

Opinions of the Press on the "International Scientific Series."

I.

Tyndall's Forms of Water.

I vol., 12mo. Cloth. Illustrated. . . . . Price, $1.50.

"In the volume now published, Professor Tyndall has presented a noble illustration of the acuteness and subtlety of his intellectual powers, the scope and insight of his scientific vision, his singular command of the appropriate language of exposition, and the peculiar vivacity and grace with which he unfolds the results of intricate scientific research."—*N. Y. Tribune.*

"The 'Forms of Water,' by Professor Tyndall, is an interesting and instructive little volume, admirably printed and illustrated. Prepared expressly for this series, it is in some measure a guarantee of the excellence of the volumes that will follow, and an indication that the publishers will spare no pains to include in the series the freshest investigations of the best scientific minds."—*Boston Journal.*

"This series is admirably commenced by this little volume from the pen of Prof. Tyndall. A perfect master of his subject, he presents in a style easy and attractive his methods of investigation, and the results obtained, and gives to the reader a clear conception of all the wondrous transformations to which water is subjected."—*Churchman.*

II.

Bagehot's Physics and Politics.

I vol., 12mo. Price, $1.50.

"If the 'International Scientific Series' proceeds as it has begun, it will more than fulfil the promise given to the reading public in its prospectus. The first volume, by Professor Tyndall, was a model of lucid and attractive scientific exposition; and now we have a second, by Mr. Walter Bagehot, which is not only very lucid and charming, but also original and suggestive in the highest degree. Nowhere since the publication of Sir Henry Maine's 'Ancient Law,' have we seen so many fruitful thoughts suggested in the course of a couple of hundred pages. . . . To do justice to Mr. Bagehot's fertile book, would require a long article. With the best of intentions, we are conscious of having given but a sorry account of it in these brief paragraphs. But we hope we have said enough to commend it to the attention of the thoughtful reader."—Prof. JOHN FISKE, in the *Atlantic Monthly.*

"Mr. Bagehot's style is clear and vigorous. We refrain from giving a fuller account of these suggestive essays, only because we are sure that our readers will find it worth their while to peruse the book for themselves; and we sincerely hope that the forthcoming parts of the 'International Scientific Series' will be as interesting."—*Athenæum.*

"Mr. Bagehot discusses an immense variety of topics connected with the progress of societies and nations, and the development of their distinctive peculiarities; and his book shows an abundance of ingenious and original thought."—ALFRED RUSSELL WALLACE, in *Nature.*

D. APPLETON & CO., Publishers, 549 & 551 Broadway, N. Y.

III.

Foods.

By Dr. EDWARD SMITH.

1 vol., 12mo. Cloth. Illustrated. . . . . Price, $1.75.

In making up THE INTERNATIONAL SCIENTIFIC SERIES, Dr. Edward Smith was selected as the ablest man in England to treat the important subject of Foods. His services were secured for the undertaking, and the little treatise he has produced shows that the choice of a writer on this subject was most fortunate, as the book is unquestionably the clearest and best-digested compend of the Science of Foods that has appeared in our language.

"The book contains a series of diagrams, displaying the effects of sleep and meals on pulsation and respiration, and of various kinds of food on respiration, which, as the results of Dr. Smith's own experiments, possess a very high value. We have not far to go in this work for occasions of favorable criticism; they occur throughout, but are perhaps most apparent in those parts of the subject with which Dr. Smith's name is especially linked."—*London Examiner.*

"The union of scientific and popular treatment in the composition of this work will afford an attraction to many readers who would have been indifferent to purely theoretical details. . . . Still his work abounds in information, much of which is of great value, and a part of which could not easily be obtained from other sources. Its interest is decidedly enhanced for students who demand both clearness and exactness of statement, by the profusion of well-executed woodcuts, diagrams, and tables, which accompany the volume. . . . The suggestions of the author on the use of tea and coffee, and of the various forms of alcohol, although perhaps not strictly of a novel character, are highly instructive, and form an interesting portion of the volume."—*N. Y. Tribune.*

IV.

Body and Mind.

THE THEORIES OF THEIR RELATION.

By ALEXANDER BAIN, LL. D.

1 vol., 12mo. Cloth. . . . . Price, $1.50.

PROFESSOR BAIN is the author of two well-known standard works upon the Science of Mind—"The Senses and the Intellect," and "The Emotions and the Will." He is one of the highest living authorities in the school which holds that there can be no sound or valid psychology unless the mind and the body are studied, as they exist, together.

"It contains a forcible statement of the connection between mind and body, studying their subtile interworkings by the light of the most recent physiological investigations. The summary in Chapter V., of the investigations of Dr. Lionel Beale of the embodiment of the intellectual functions in the cerebral system, will be found the freshest and most interesting part of his book. Prof. Bain's own theory of the connection between the mental and the bodily part in man is stated by himself to be as follows: There is 'one substance, with two sets of properties, two sides, the physical and the mental—a *double-faced unity.*' While, in the strongest manner, asserting the union of mind with brain, he yet denies 'the association of union *in place,*' but asserts the union of close succession in time,' holding that 'the same being is, by alternate fits, under extended and under unextended consciousness." '—*Christian Register.*

D. APPLETON & CO., Publishers, 549 & 551 Broadway, N. Y.

V.

The Study of Sociology.

By HERBERT SPENCER.

1 vol., 12mo. Cloth. . . . . Price, $1.50.

"The philosopher whose distinguished name gives weight and influence to this volume, has given in its pages some of the finest specimens of reasoning in all its forms and departments. There is a fascination in his array of facts, incidents, and opinions, which draws on the reader to ascertain his conclusions. The coolness and calmness of his treatment of acknowledged difficulties and grave objections to his theories win for him a close attention and sustained effort, on the part of the reader, to comprehend, follow, grasp, and appropriate his principles. This book, independently of its bearing upon sociology, is valuable as lucidly showing what those essential characteristics are which entitle any arrangement and connection of facts and deductions to be called a *science*."—*Episcopalian.*

"This work compels admiration by the evidence which it gives of immense research, study, and observation, and is, withal, written in a popular and very pleasing style. It is a fascinating work, as well as one of deep practical thought."—*Bost. Post.*

"Herbert Spencer is unquestionably the foremost living thinker in the psychological and sociological fields, and this volume is an important contribution to the science of which it treats. . . . It will prove more popular than any of its author's other creations, for it is more plainly addressed to the people and has a more practical and less speculative cast. It will require thought, but it is well worth thinking about."—*Albany Evening Journal.*

VI.

The New Chemistry.

By JOSIAH P. COOKE, JR.,

Erving Professor of Chemistry and Mineralogy in Harvard University.

1 vol., 12mo. Cloth. . . . . Price, $2.00.

"The book of Prof. Cooke is a model of the modern popular science work. It has just the due proportion of fact, philosophy, and true romance, to make it a fascinating companion, either for the voyage or the study."—*Daily Graphic.*

"This admirable monograph, by the distinguished Erving Professor of Chemistry in Harvard University, is the first American contribution to 'The International Scientific Series,' and a more attractive piece of work in the way of popular exposition upon a difficult subject has not appeared in a long time. It not only well sustains the character of the volumes with which it is associated, but its reproduction in European countries will be an honor to American science."—*New York Tribune.*

"All the chemists in the country will enjoy its perusal, and many will seize upon it as a thing longed for. For, to those advanced students who have kept well abreast of the chemical tide, it offers a calm philosophy. To those others, youngest of the class, who have emerged from the schools since new methods have prevailed, it presents a generalization, drawing to its use all the data, the relations of which the newly-fledged fact-seeker may but dimly perceive without its aid. . . . To the old chemists, Prof. Cooke's treatise is like a message from beyond the mountain. They have heard of changes in the science; the clash of the battle of old and new theories has stirred them from afar. The tidings, too, had come that the old had given way; and little more than this they knew. . . . Prof. Cooke's 'New Chemistry' must do wide service in bringing to close sight the little known and the longed for. . . . As a philosophy it is elementary, but, as a book of science, ordinary readers will find it sufficiently advanced."—*Utica Morning Herald.*

VII.

The Conservation of Energy.

By BALFOUR STEWART, LL. D., F. R. S.

With an Appendix treating of the Vital and Mental Applications of the Doctrine.

1 vol., 12mo. Cloth. Price, $1.50.

"The author has succeeded in presenting the facts in a clear and satisfactory manner, using simple language and copious illustration in the presentation of facts and principles, confining himself, however, to the physical aspect of the subject. In the Appendix the operation of the principles in the spheres of life and mind is supplied by the essays of Professors Le Conte and Bain."—*Ohio Farmer.*

"Prof. Stewart is one of the best known teachers in Owens College in Manchester.

"The volume of THE INTERNATIONAL SCIENTIFIC SERIES now before us is an excellent illustration of the true method of teaching, and will well compare with Prof. Tyndall's charming little book in the same series on 'Forms of Water,' with illustrations enough to make clear, but not to conceal his thoughts, in a style simple and brief."—*Christian Register, Boston.*

"The writer has wonderful ability to compress much information into a few words. It is a rich treat to read such a book as this, when there is so much beauty and force combined with such simplicity.—*Eastern Press.*

VIII.

Animal Locomotion;

Or, WALKING, SWIMMING, AND FLYING.

With a Dissertation on Aëronautics.

By J. BELL PETTIGREW, M. D., F. R. S., F. R. S. E., F. R. C. P. E.

1 vol., 12mo. . . . . . . Price, $1.75.

"This work is more than a contribution to the stock of entertaining knowledge, though, if it only pleased, that would be sufficient excuse for its publication. But Dr. Pettigrew has given his time to these investigations with the ultimate purpose of solving the difficult problem of Aëronautics. To this he devotes the last fifty pages of his book. Dr. Pettigrew is confident that man will yet conquer the domain of the air."—*N. Y. Journal of Commerce.*

"Most persons claim to know how to walk, but few could explain the mechanical principles involved in this most ordinary transaction, and will be surprised that the movements of bipeds and quadrupeds, the darting and rushing motion of fish, and the erratic flight of the denizens of the air, are not only anologous, but can be reduced to similar formula. The work is profusely illustrated, and, without reference to the theory it is designed to expound, will be regarded as a valuable addition to natural history." —*Omaha Republic.*

IX.

Responsibility in Mental Disease.

By HENRY MAUDSLEY, M. D.,

Fellow of the Royal College of Physicians; Professor of Medical Jurisprudence in University College, London.

1 vol., 12mo. Cloth. . . Price, $1.50.

"Having lectured in a medical college on Mental Disease, this book has been a feast to us. It handles a great subject in a masterly manner, and, in our judgment, the positions taken by the author are correct and well sustained."—*Pastor and People.*

"The author is at home in his subject, and presents his views in an almost singularly clear and satisfactory manner. . . . The volume is a valuable contribution to one of the most difficult, and at the same time one of the most important subjects of investigation at the present day."—*N. Y. Observer.*

"It is a work profound and searching, and abounds in wisdom."—*Pittsburg Commercial.*

"Handles the important topic with masterly power, and its suggestions are practical and of great value."—*Providence Press.*

X.

The Science of Law.

By SHELDON AMOS, M. A.,

Professor of Jurisprudence in University College, London; author of "A Systematic View of the Science of Jurisprudence," "An English Code, its Difficulties and the Modes of overcoming them," etc., etc.

1 vol., 12mo. Cloth. . . . . Price, $1.75.

"The valuable series of 'International Scientific' works, prepared by eminent specialists, with the intention of popularizing information in their several branches of knowledge, has received a good accession in this compact and thoughtful volume. It is a difficult task to give the outlines of a complete theory of law in a portable volume, which he who runs may read, and probably Professor Amos himself would be the last to claim that he has perfectly succeeded in doing this. But he has certainly done much to clear the science of law from the technical obscurities which darken it to minds which have had no legal training, and to make clear to his 'lay' readers in how true and high a sense it can assert its right to be considered a science, and not a mere practice."—*The Christian Register.*

"The works of Bentham and Austin are abstruse and philosophical, and Maine's require hard study and a certain amount of special training. The writers also pursue different lines of investigation, and can only be regarded as comprehensive in the departments they confined themselves to. It was left to Amos to gather up the result and present the science in its fullness. The unquestionable merits of this, his last book, are, that it contains a complete treatment of a subject which has hitherto been handled by specialists, and it opens up that subject to every inquiring mind. . . . To do justice to 'The Science of Law' would require a longer review than we have space for. We have read no more interesting and instructive book for some time. Its themes concern every one who renders obedience to laws, and who would have those laws the best possible. The tide of legal reform which set in fifty years ago has to sweep yet higher if the flaws in our jurisprudence are to be removed. The process of change cannot be better guided than by a well-informed public mind, and Prof. Amos has done great service in materially helping to promote this end."—*Buffalo Courier.*

D. APPLETON & CO., PUBLISHERS, 549 & 551 Broadway, N. Y.

XI.

Animal Mechanism,

A Treatise on Terrestrial and Aërial Locomotion.

By E. J. MAREY,

Professor at the College of France, and Member of the Academy of Medicine.

With 117 Illustrations, drawn and engraved under the direction of the author.

1 vol., 12mo. Cloth. . . . . Price, $1.75

"We hope that, in the short glance which we have taken of some of the most important points discussed in the work before us, we have succeeded in interesting our readers sufficiently in its contents to make them curious to learn more of its subject-matter. We cordially recommend it to their attention.

"The author of the present work, it is well known, stands at the head of those physiologists who have investigated the mechanism of animal dynamics—indeed, we may almost say that he has made the subject his own. By the originality of his conceptions, the ingenuity of his constructions, the skill of his analysis, and the perseverance of his investigations, he has surpassed all others in the power of unveiling the complex and intricate movements of animated beings."—*Popular Science Monthly.*

XII.

History of the Conflict between Religion and Science.

By JOHN WILLIAM DRAPER, M. D., LL. D.,

Author of "The Intellectual Development of Europe."

1 vol., 12mo. . . . . Price, $1.75.

"This little 'History' would have been a valuable contribution to literature at any time, and is, in fact, an admirable text-book upon a subject that is at present engrossing the attention of a large number of the most serious-minded people, and it is no small compliment to the sagacity of its distinguished author that he has so well gauged the requirements of the times, and so adequately met them by the preparation of this volume. It remains to be added that, while the writer has flinched from no responsibility in his statements, and has written with entire fidelity to the demands of truth and justice, there is not a word in his book that can give offense to candid and fair-minded readers."—*N. Y. Evening Post.*

"The key-note to this volume is found in the antagonism between the progressive tendencies of the human mind and the pretensions of ecclesiastical authority, as developed in the history of modern science. No previous writer has treated the subject from this point of view, and the present monograph will be found to possess no less originality of conception than vigor of reasoning and wealth of erudition. . . . The method of Dr. Draper, in his treatment of the various questions that come up for discussion, is marked by singular impartiality as well as consummate ability. Throughout his work he maintains the position of an historian, not of an advocate. His tone is tranquil and serene, as becomes the search after truth, with no trace of the impassioned ardor of controversy. He endeavors so far to identify himself with the contending parties as to gain a clear comprehension of their motives, but, at the same time, he submits their actions to the tests of a cool and impartial examination."—*N. Y. Tribune.*

RECENT PUBLICATIONS.

THE NATIVE RACES OF THE PACIFIC STATES.

By HERBERT H. BANCROFT. To be completed in 5 vols. Vol. I. now ready. Containing Wild Tribes: their Manners and Customs. 1 vol., 8vo. Cloth, $6; sheep, $7.

"We can only say that if the remaining volumes are executed in the same spirit of candid and careful investigation, the same untiring industry, and intelligent good sense, which mark the volume before us, Mr. Bancroft's 'Native Races of the Pacific States will form, as regards aboriginal America, an encyclopædia of knowledge not only unequaled but unapproached. A literary enterprise more deserving of a generous sympathy and support has never been undertaken on this side of the Atlantic."—FRANCIS PARKMAN, in the *North American Review.*

"The industry, sound judgment, and the excellent literary style displayed in this work, cannot be too highly praised."—*Boston Post.*

A BRIEF HISTORY OF CULTURE.

By JOHN S. HITTELL. 1 vol., 12mo. Price, $1.50.

"He writes in a popular style for popular use. He takes ground which has never been fully occupied before, although the general subject has been treated more or less distinctly by several writers. . . . Mr. Hittell's method is compact, embracing a wide field in a few words, often presenting a mere hint, when a fuller treatment is craved by the reader; but, although his book cannot be commended as a model of literary art, it may be consulted to great advantage by every lover of free thought and novel suggestions."—*N. Y. Tribune.*

THE HISTORY OF THE CONFLICT BETWEEN RELIGION AND SCIENCE.

By JOHN W. DRAPER, M. D., author of "The Intellectual Development of Europe." 1 vol., 12mo. Cloth. Price, $1.75.

"The conflict of which he treats has been a mighty tragedy of humanity that has dragged nations into its vortex and involved the fate of empires. The work, though small, is full of instruction regarding the rise of the great ideas of science and philosophy; and he describes in an impressive manner and with dramatic effect the way religious authority has employed the secular power to obstruct the progress of knowledge and crush out the spirit of investigation. While there is not in his book a word of disrespect for things sacred, he writes with a directness of speech, and a vividness of characterization and an unflinching fidelity to the facts, which show him to be in thorough earnest with his work. The 'History of the Conflict between Religion and Science' is a fitting sequel to the 'History of the Intellectual Development of Europe,' and will add to its author's already high reputation as a philosophic historian."—*N. Y. Tribune.*

THEOLOGY IN THE ENGLISH POETS.

COWPER, COLERIDGE, WORDSWORTH, and BURNS. By Rev. STOPFORD BROOKE. 1 vol., 12mo. Price, $2.

"Apart from its literary merits, the book may be said to possess an independent value, as tending to familiarize a certain section of the English public with more enlightened views of theology."—*London Athenæum.*

BLOOMER'S COMMERCIAL CRYPTOGRAPH.

A Telegraph Code and Double Index—Holocryptic Cipher. By J. G. BLOOMER. 1 vol., 8vo. Price, $5.

By the use of this work, business communications of whatever nature may be telegraphed with secrecy and economy.

D. APPLETON & CO., Publishers, New York.

www.ingramcontent.com/pod-product-compliance
Lightning Source LLC
LaVergne TN
LVHW011206110826
845150LV00006B/1342

* 9 7 8 1 4 2 5 5 1 8 0 1 1 *